ILLUSTRATED ADVANCED BIOLOGY

Mammals: Structure & Function

C J Clegg

JOHN MURRAY

Also in this series:

Genetics and Evolution 0 7195 7552 4

Artwork credits

The following are sources from which artwork or data have been adapted and redrawn with permission:

Figure 3.17 (p.20) from Fig. 101 on p.239 of J.Z. Young (1957) *The Life of Mammals*, by permission of Oxford University Press; **Figure 3.23** (p.23) from Fig. 37.17 on p.807 of *Biology* (2nd Edition) by Neil A. Campbell, © 1990, by permission of Benjamin Cummings Publishing Company; **Figure 4.10** (p.28) from *Principles of Anatomy and Physiology* (5th Edition), by Gerald Tortora and Nicholas P. Anagnostakos, © 1988 by Biological Sciences Textbooks Inc, A and P Textbooks, and Ella-Sparta. By permission of Addison-Wesley Educational Publishers; **Figure 6.2** (p.43) from Fig. 18.15 on p.387 of C.J. Clegg with D.G. Mackean (1994) *Advanced Biology: Principles and Applications* by John Murray; **Figure 7.2** (p.48) from Fig. 23.6 on p.514 of C.J. Clegg with D.G. Mackean (1994) *Advanced Biology: Principles and Applications* by John Murray; **Figure 7.13** (p.55) from original illustration by Guy Richardson on p.42 of P. Richardson (1985) *Bats* by permission of Whittet Books; **Figure 7.15** (p.56) from I.L. Mason (1984) *Evolution of Domesticated Animals.* By permission of Addison-Wesley Longman Ltd; **Figure 7.16** (p.56) from Fig. 8.20 on p.281 of K. Schmidt-Nielsen (1983) *Animal Physiology: Adaptation & Environment* (3rd Edition), by permission of Cambridge University Press; **Figure 7.18** (p.57) from Table 9.13 on p.351 of K. Schmidt-Nielsen (1983) *Animal Physiology: Adaptation & Environment* (3rd Edition), by permission of Cambridge University Press; **Figure 8.3 part** (p.59) from Fig. 10.21 on p.217 of C.J. Clegg with D.G. Mackean (1994) *Advanced Biology: Principles and Applications* by John Murray; **Figure 8.19** (p.68) from Fig. 21.29 on p.466 of C.J. Clegg with D.G.

Mackean (1994) *Advanced Biology: Principles and Applications* by John Murray; **Figure 10.3 part** (p.79) from p.126 of H.G.Q. Rowett (1988) *Basic Anatomy and Physiology* (3rd Edition) by John Murray; **Figure 10.8** (p.81) from p.93 of W.H. Freeman & B. Bracegirdle (1966) *An Atlas of Histology* by permission of Heinemann Educational Publishers, a division of Reed Educational & Professional Publishing Ltd; **Figure 10.12** (p.83) from part of Fig. 7.1 on p.129 of J. Cohen (1977) *Reproduction* by Butterworth.

Photo credits

All photographs are reproduced with permission of the copyright holders. The bulk of the photomicrographs are the property of Gene Cox. The others, identified in the listing below, are the property of:

Cover © Gene Cox; **p.2** Chris Clegg; **p.3** *top* © Niall Benvie/BBC Natural History Unit, *bottom* © Armin Maywald/BBC Natural History Unit; **p.5** Biophoto Associates; **p.7** Dr Gopal Murti/Science Photo Library; **p.8** Don Fawcett/Science Photo Library; **p.14** Chris Clegg; **p.15** *top right* Biophoto Associates; **p.22** Biophoto Associates; **p.25** Chris Clegg; **p.29** *top left & top right* Biophoto Associates, *bottom* National Medical Slide Bank; **p.35** Prof. P. Motta/Dept of Anatomy/University "La Sapienza", Rome/Science Photo Library; **p.39** *top* CNRI/Science Photo Library, *bottom* Moredun Animal Health Ltd/Science Photo Library; **p.53** Biophoto Associates; **p.55** © Stephen Dalton/NHPA; **p.56** Chris Clegg; **p.57** *top left & top right* Chris Clegg, *bottom* © Mary Ann Macdonald/BBC Natural History Unit; **p.62** Pr S. Cinti/CNRI/Science Photo Library; **p.74** *bottom* © Life Science Images; **p.83** Petit Format/CSI/Science Photo Library.

First published in 1998
by John Murray (Publishers) Ltd
50 Albemarle Street
London W1X 4BD

Illustrations by C J Clegg
with additional illustrations by Mike Humphries and Lydia Umney

Layouts by Fiona Webb
Cover design by John Townson/Creation

Typeset in 10/12pt Galliard by Wearset, Boldon, Tyne and Wear
Printed and bound in Great Britain by the Alden Group, Oxford

A catalogue entry for this title is available from the British Library

ISBN 0 7195 7551 6

Contents

Abbreviations used in captions

TEM	transmission electronmicrograph
SEM	scanning electronmicrograph
HP	high power
MP	medium power
LP	low power
TS	transverse section
LS	longitudinal section

Preface

This book is one of a series that is designed to support active learning. Here the concepts of 'function' are linked to the investigation of structure in mammals, in a pictorial format.

Mammals are a comparatively recent group in terms of their evolutionary history, yet they have successfully settled in significant numbers in virtually every type of habitat on the Earth's surface. This contrasts with the insects, for example, a very ancient class of animals by comparison (and numerically vastly more successful), yet they have been unable to master the full range of habitats in which mammals have excelled.

The success of the mammals is undoubtedly linked to mastery of their own internal environment: to the efficiency with which they extract nutrients in the gut; circulate and supply essential nutrients to tissues; move and maintain posture; accurately communicate data on their internal and external conditions; and subtly integrate their body systems. All these facets to success are treated in appropriate depth, using photomicrographs, electronmicrographs and photographs, which are interpreted by drawings, flow-diagrams, graphs and notes.

The treatment is ideal for students at AS, A level and GNVQ in Biology; for students of Environmental Studies, Social and Human Biology, and all related disciplines as well as being of general interest.

Taking your studies further

- You can read more about the background to these topics, and learn how they have been investigated in the laboratory and in nature from: CJ Clegg with DG Mackean (1994) *Advanced Biology Principles & Applications* (by John Murray).
- Related laboratory practicals, projects and investigations, and details of other resources for learning are available in: CJ Clegg with DG Mackean, PH Openshaw and RC Reynolds (1996) *Advanced Biology Study Guide* (by John Murray).
- Keeping up-to-date with developments in mammalian classification, ecology and physiology, and in medicine may come from articles in:
  New Scientist, a weekly review of developments, with the excellent occasional series 'Inside Science';
  Biological Sciences Review, a relatively new journal, designed and written for A level biology students;
  Scientific American, an excellent review journal of science, sometimes with articles applicable to this level.

Acknowledgements

To all the known and unknown scientists, naturalists, teachers, illustrators and writers who have influenced my own understanding I gladly acknowledge my debt. Where copyright material has been used it is acknowledged on page ii.

If the intellectual property of anyone has inadvertently been used without prior agreement, then I ask that John Murray (Publishers) Ltd, of 50 Albemarle Street, London W1X 4BD are contacted so that correction can be made.

I am greatly indebted to friends and colleagues, including Don Mackean, for opportunities to review my ideas and the presentation of them in text and illustrations, and for their many useful suggestions. The remaining errors are my sole responsibility.

At John Murray the skill and patience of Katie Mackenzie Stuart, Helen Townson and Julie Jones have brought together text, photomicrographs and drawings exactly as I wished, and I am most grateful to them.

C J Clegg
September 1998

Introduction to classification and diversity

There are vast numbers of living things. Some estimates suggest that 100 million or more different species exist today. Biologists use a classification system to impose an order and general plan upon living things. The system used imposes two classifications upon organisms:

1 An **internationally agreed name**: in the binomial system a name consists of two words, in Latin. The first name (a noun) is the **genus** (e.g. *Canis*, for wolves and dogs); the second (an adjective) the **species** (e.g. *familiaris*, the domesticated dog). The genus is shared with related species, e.g. *Canis lupus*, the wolf.

   We are now confident that living things change with time, and that species have evolved, one from another. So, we say that a species consists of *organisms of common ancestry that closely resemble each other structurally and biochemically, and that are members of a natural population that are actually or potentially capable of breeding with each other to produce fertile progeny, and that do not interbreed with members of other species.*

2 A **hierarchical plan**: species are placed in groupings that in turn are gathered into larger groups. The taxa (groupings) used are:

$$\textbf{kingdom} \rightarrow \textbf{phylum} \rightarrow \textbf{class} \rightarrow \textbf{order} \rightarrow \textbf{family} \rightarrow \textbf{genus} \rightarrow \textbf{species}$$

It turns out that some taxa are inconveniently large. This is overcome by the use of subdivisions, like 'sub-phylum'.

As many characteristics as possible are used to classify similar organisms together: differences in structure and biochemistry of organisms, details of the ultrastructure of the cells of organisms, evolutionary relationships as far as these are known, all contribute to the **five-kingdom scheme of classification** currently preferred.

Figure 1.1 Classification of mammals within the five-kingdom scheme.
Electronmicroscopy has shown a fundamental difference in cellular organisation between **prokaryotes** (have small cells and no true nucleus) and **eukaryotes** (have a larger cell and a true nucleus).

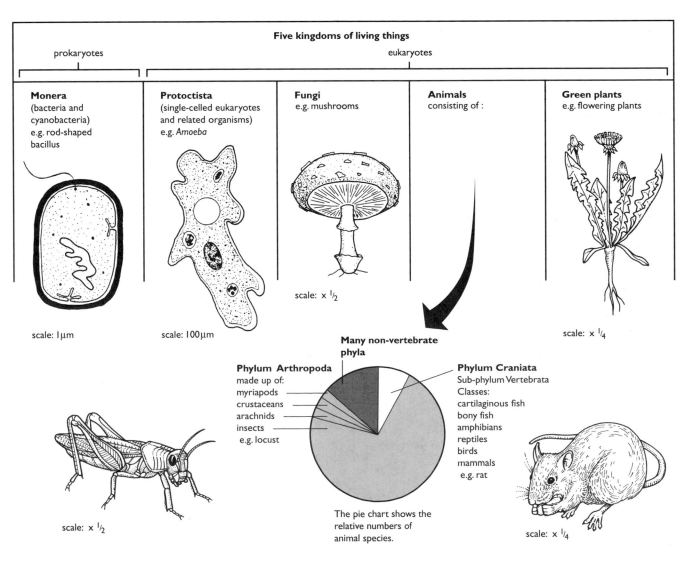

Five kingdoms of living things

prokaryotes — eukaryotes

Monera
(bacteria and cyanobacteria)
e.g. rod-shaped bacillus

scale: 1µm

Protoctista
(single-celled eukaryotes and related organisms)
e.g. *Amoeba*

scale: 100µm

Fungi
e.g. mushrooms

scale: x ½

Animals
consisting of :

Green plants
e.g. flowering plants

scale: x ¼

Many non-vertebrate phyla

Phylum Arthropoda
made up of:
myriapods
crustaceans
arachnids
insects
  e.g. locust

scale: x ½

The pie chart shows the relative numbers of animal species.

Phylum Craniata
Sub-phylum Vertebrata
Classes:
cartilaginous fish
bony fish
amphibians
reptiles
birds
mammals
  e.g. rat

scale: x ¼

Taxonomy of a common mammal

Kingdom: Animalia (animals)

- Multicellular, eukaryotic (with true nucleus) organisms.
- Heterotrophic nutrition (incapable of manufacturing food, dependent on other organisms).
- Typically have cells organised into tissues and organs.
- Typically have a nervous system to coordinate actions and responses.
- In sexual reproduction animals produce haploid gametes (sperms + ova). The zygote divides repeatedly and the new cells form a hollow ball, the blastula.

Phylum: Chordata (chordates)

Show the following five characteristics, at least at some stage during development.

- Tubular, hollow, dorsal nerve cord.
- Dorsal, flexible supporting rod called the notochord.
- Post-anal tail.
- Set of gill slits in the throat, the pharyngeal or visceral clefts.
- Blood circulation system in which blood flows down the body dorsally and returns ventrally.

Figure 1.2 The domesticated dog (*Canis familiaris*) was selectively bred (artificial selection) from the wolf (*Canis lupus*) in Neolithic times, over 9500 years ago.

Sub-phylum: Vertebrata (vertebrates or craniates)

- The notochord is replaced by a vertebral column (backbone).
- A brain is present and is enclosed in a cranium.

Class: Mammalia (mammals)

- Characterised by their great activity, high rate of metabolism, 'intelligence' and their high degree of parental care (shown also by birds).
- Skin covered by hair, and typically contains two types of glands, sweat glands and sebaceous glands.
- Body divided by muscular diaphragm between thorax and abdomen.
- Body maintained at relatively high and constant temperature by internal regulation (endothermic).
- Fertilisation is internal; after birth the young are fed on milk from the mammary glands.

Sub-class: Eutheria (placental mammals)

- Young initially develop in the uterus and are nurtured by the maternal blood circulation via the placenta.

Order: Carnivora (carnivores)

- Well equipped for meat eating: powerful jaws, large canine teeth, 'cheek' teeth for shearing, and claws.
- Many have agile, graceful bodies (typically the cats); others are more heavily built (typically the bears).

Family: Canidae (wolves, jackals, foxes)

- Long slender limbs and bushy tails. Generally good runners, running on the tips of toes (four toes on the front feet).
- Keen hearing and sight, but hunt by scent, often running in packs.

e.g. Genus + species: *Canis familiaris*

1 Which of the external features shown by a panting dog help to identify its phylum, class and order?

Range of life within the mammals

There are more orders within the class Mammalia than are listed here, but this selection gives an introduction. *Note:* the numbers of species within the (non-vertebrate) insects far out-number those of mammals. However, mammals have mastered virtually every sort of habitat, including the sea (which makes up three-quarters of the Earth's surface) where insects are not found at all.

Figure 1.3 Orders of true mammals.

Insectivora e.g. shrews, hedgehogs and moles
Small mammals, mostly nocturnal, with long, sensitive snouts. Biologists believe that these mammals are a primitive group, with features little changed from those of the earliest eutherian mammals whose fossil forms are found in rocks of the Cretaceous, approximately 135 million years ago

Rodentia e.g. squirrels, beavers, rats and mice
The most successful modern mammals apart from humans, being found in all parts of the world

Lagomorpha e.g. hares and rabbits
These animals resemble rodents, with a second pair of upper incisors

Carnivora e.g. cats, dogs, weasels, bears (and pandas)
Mostly flesh-eating mammals with powerful jaws and large canine teeth

Artiodactyla e.g. cattle, sheep, pigs, giraffe, deer and camels
Also known as the 'even-toed, hoofed mammals'

Hedgehog (*Erinaceus europaeus*)

Dolphin (*Delphinus delphis*)

Cetacea e.g. whales and dolphins
Aquatic, fish-like mammals that have mastered life in the sea

Perissodactyla e.g. rhinoceroses, horses and zebras
Also known as the 'odd-toed, hoofed mammals'

Proboscidea e.g. elephants
Have a unique extension of the nose as a flexible trunk

Chiroptera e.g. bats
Small mammals that are fully adapted for flight

Primates: e.g. tree-shrews, lemurs, monkeys, apes and humans
Mostly tree-dwelling mammals; limbs with five digits and with grasping hands and feet. Eyes are at the front of the head, close together and directed forwards, providing stereoscopic vision. Apart from humans, most are tropical or sub-tropical species

Other mammals

The majority of living mammals belong to one of the orders previously described, or to similar orders. However, mammals appear to have evolved independently at least three times. We find two other orders that, whilst having many mammal-like features, are clearly distinct from eutherian (placental) mammals. These orders are largely restricted to Australia and South America.

- **Order Monotremata: the egg-laying mammals.** Example: duck-billed platypus (*Ornithorhynchus* sp.).
   Here the reproductive system is reminiscent of reptiles, and eggs are laid, which are then incubated. However, they are endothermic animals, they have hair and they suckle their young.
- **Order Marsupialia: the pouched mammals.** Example: kangaroo (*Macropus* sp.).
   These animals have more similarities with eutherian mammals than do the monotremes. They are a more diverse group, paralleling some of the different types of animal found among eutherian mammals. In contrast, however, they have a very short gestation period and young are born at an early stage of development. The young crawl through the fur to a pouch on the mother's body, where they attach themselves to a nipple. Most development occurs here.

2 Cells and tissues

A **cell**, the basic unit of life, is an extremely small structure. In multicellular organisms like mammals there are many millions of cells present, organised into tissues and organs. A **tissue** is a group of cells of similar structure that perform a particular function, for example cardiac muscle tissue. An **organ** is part of an organism, made of a collection of tissues, for example the heart (cardiac muscle and connective tissue).

An animal cell consists of a **nucleus** surrounded by **cytoplasm**, contained within a cell membrane (**plasma membrane**). The nucleus controls and directs the activity of the cell. Inside the nucleus are **chromosomes**, which appear dispersed as chromatin, except during nuclear divisions. The cytoplasm is the site of the chemical reactions of life (known as **metabolism**).

1 Originally stated, the cell theory recognised that organisms consist of cells and cell products. Today, the cell theory includes four concepts. What are they?

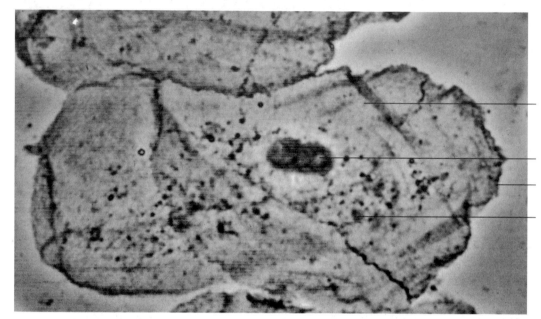

Figure 2.1 Human cheek cell observed under HP by bright-field microscopy (×2500).

cytoplasm

nucleus

plasma membrane

organelle in cytoplasm

Cell division

When a cell divides the nucleus divides first, followed by the cytoplasm. In the first step of nuclear division the chromosomes are precisely copied. Then, in **mitosis**, two daughter cells are formed containing identical sets of chromosomes. In **meiosis**, four daughter nuclei are formed, each containing half the chromosome number.

Figure 2.2 Cell division by mitosis and meiosis.

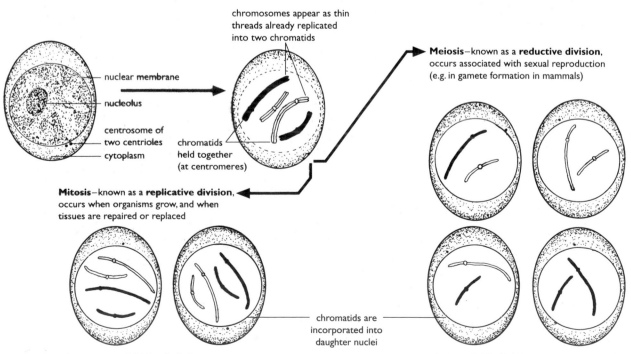

chromosomes appear as thin threads already replicated into two chromatids

nuclear membrane

nucleolus

centrosome of two centrioles

cytoplasm

chromatids held together (at centromeres)

Meiosis–known as a **reductive division**, occurs associated with sexual reproduction (e.g. in gamete formation in mammals)

Mitosis–known as a **replicative division**, occurs when organisms grow, and when tissues are repaired or replaced

chromatids are incorporated into daughter nuclei

two nuclei, identical chromosomes

four nuclei, half the chromosome number

Observing cells and tissues

Cells are examined with the aid of a microscope, often by **optical microscopy** using visible light, or by **electronmicroscopy** using a beam of electrons. **Resolution** of a microscope, its ability to separate small objects very close together, is related to the wavelength of illumination used. The wavelength of visible light is of the order of 500 nm, whereas the wavelength of the electron beam used is 0.005 nm. Consequently, whilst a light microscope can, at best, resolve structures 200 nm (0.2 μm) apart, the transmission electronmicroscope can resolve points 1nm apart when used on biological specimens.

Units of length used in microscopy:

1 metre (m) = 1000 millimetres (mm); 1 mm (10^{-3} m) = 1000 **micrometres (μm)**
1 μm (10^{-6} m) = 1000 **nanometres (nm)**; 1 nm (10^{-9} m) = 1000 **picometres (pm)**

Much of what we know about the structure of the tissues and organs of mammals has been learnt by microscopic observations of tissues that have been fixed (killed and preserved in a lifelike condition), sectioned, stained and mounted on microscope slides, and then examined by **normal bright-field light microscopy**. The image of an animal cell opposite is an example. Alternatively, in phase contrast microscopy, structures in unstained, living cells can be observed in high contrast with good resolution. A photomicrograph taken by **phase contrast microscopy** is shown in Figure 9.4 (page 74).

Investigation of the fine structure of cells, known as the **ultrastructure**, was made possible by the technique of **transmission electronmicroscopy** (TEM). The electron beam travels at high speed but with low energy, so the interior of the microscope must be under a vacuum. Because of this vacuum, specimens must be dehydrated and only very thin sections or tiny objects can be investigated. The cells have to be killed and chemically 'fixed' in a lifelike condition, embedded in a supporting resin, and sectioned to less than 2 μm thickness. These sections are then stained with electron-dense chemicals. The dried, stained sections are introduced into the electronmicroscope for examination.

Figure 2.3 TEM of a liver cell, HP.

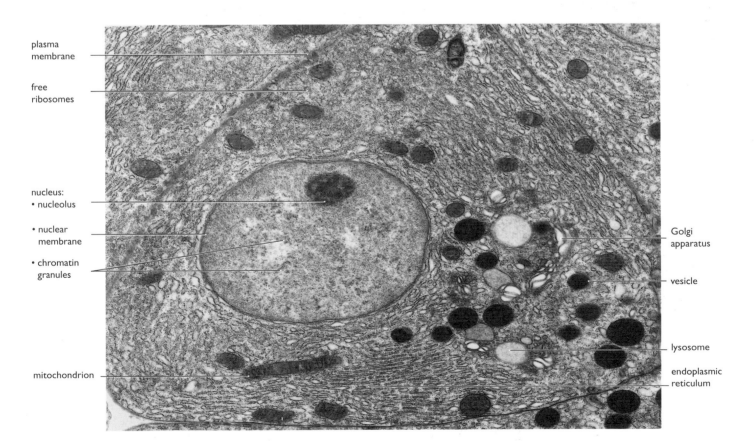

In a **scanning electronmicroscope** the specimen is scanned by a beam of electrons and an image is created from the electrons reflected from the specimen's surface. Larger specimens can be viewed than by TEM, but the resolution is not as good. A scanning electronmicrograph (SEM) is shown in Figure 2.6 (page 8).

Organelles of mammalian cells

The cell is a sac of **organelles** suspended within an aqueous medium, the **cytosol**, which contains essential ions and soluble organic compounds (amino acids, sugars, fatty acids and proteins). Most of the organelles are made up of membranes, similar in structure to the plasma membrane. Many membrane proteins and the free proteins in the cytosol are enzymes.

2 What is an enzyme, and (concisely) how does it 'work'?

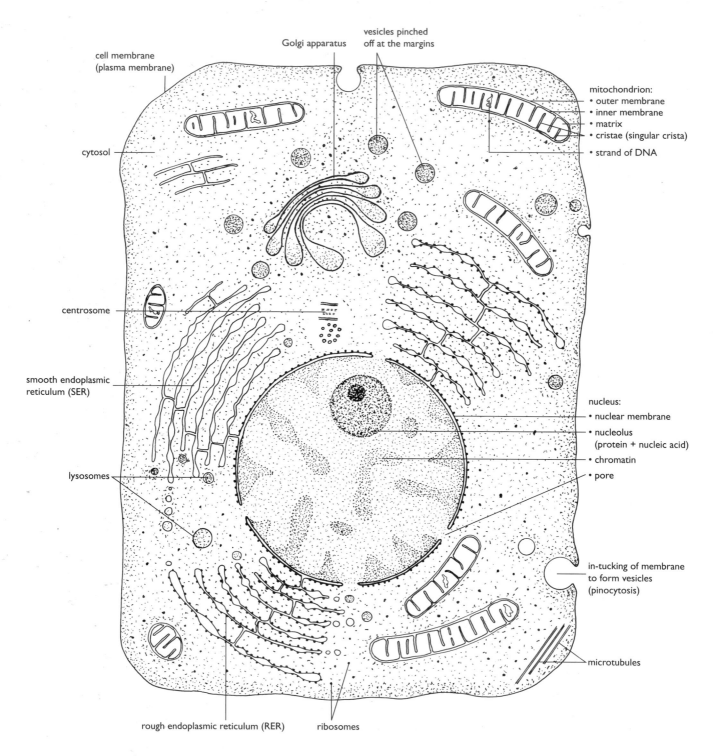

Figure 2.4 The ultrastructure of the eukaryotic cell.

- **Nucleus**: the largest organelle, typically 10–20 μm in diameter. It contains the chromosomes, visible at times of nuclear division only, otherwise dispersed as chromatin. Chromosomes contain the genetic material DNA (deoxyribonucleic acid). They control and direct the development and activities of the cell by providing information for protein/enzyme synthesis in the cytoplasm.

- **Mitochondria**: rod-shaped or cylindrical organelles that occur in all cells, usually in very large numbers. Part of aerobic respiration is located in the mitochondrion, along with some other enzyme systems. Mitochondria are variable in size, but typically 1 × 7 μm.

- **Endoplasmic reticulum** (ER): a folded membrane in the form of interconnected sheets (or tubes or sacs). Rough ER has ribosomes attached, and is the site of 'packaged' protein synthesis (e.g. digestive enzymes are prepared for export from the cell). Smooth ER lacks ribosomes, and is the site of cell lipid metabolism.

- **Ribosomes**: small, almost spherical organelles not made of membrane. They consist of protein and RNA (ribonucleic acid). RNA in this context is a rare example of an enzyme not made of protein. Ribosomes in the cytosol are the sites of synthesis of proteins that remain in the cell. Ribosomes in prokaryotes (e.g. bacteria) are smaller than those in eukaryotes.

- **Golgi apparatus**: stack-like collection of flattened membranous sacs, prominent in metabolically active cells. It is the site of synthesis of specific biochemicals (mostly enzymes, it seems) that are then packaged into vesicles.

- **Lysosomes**: small spherical packages bound by a single membrane. They are formed by a vesicle cutting off from the Golgi apparatus or the rough ER. Lysosomes contain a concentrated mixture of hydrolytic enzymes, and are typically used in the digestion and dissolution of redundant structures and damaged macromolecules in cells. Lysosomes fuse with imported food vacuoles.

- **Microtubules**: straight, unbranched hollow cylinders, made of protein. They are 25 nm wide. The tubes are built up and broken down, probably according to the needs of the cell. Microtubules are involved in the movement of cell components within the cytoplasm. They make up the centrioles of the centrosome and the spindle fibres that move chromatids in mitosis and meiosis.

- **Cilia and flagella**: organelles that project from the surface of certain cells, but are connected to a basal body just below the membrane surface.

  Cilia occur in large numbers on large cells (see Figure 2.12, page 11), whereas flagella occur singly or in small numbers typically on small, motile cells. Both consist of an outer ring of pairs of microtubules surrounding a single central pair (a 9 + 2 pattern), all enclosed in an extension of the plasma membrane. Protein side arms connect the outer pairs to the central pair and also to each other. Enzymes that release energy from ATP (adenosine triphosphate) are present, and side arm microtubules appear to work in a similar way to the 'sliding-filament' mechanism of actin and myosin in muscle myofibrils (Figure 9.7, page 75).

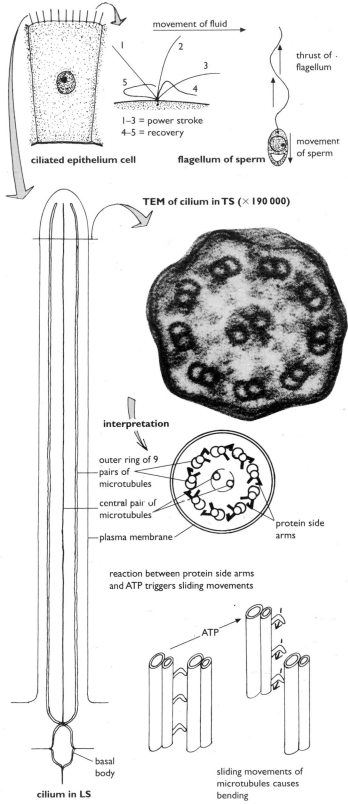

Figure 2.5 Cilia and flagella.

3 In what ways does the structure of a bacterial cell (a prokaryotic organism) differ from a mammalian cell?

The plasma membrane

The membrane around the cell acts as a barrier. It allows water, energy, nutrients and waste substances to cross it. Proteins and lipids, including phospholipids, make up the membrane. Some of these components have large carbohydrate molecules forming complexes on the exterior surface, referred to as the glycocalyx. Receptor molecules in the membrane allow the cell to 'identify' specific hormones (Figure 8.24, page 71) and other cells.

The membrane is visualised as a **'fluid mosaic model'**: the lipids are arranged in a bilayer with their non-polar tails facing inwards and polar heads outwards. The protein components are globular and occur as huge, discrete molecules on, in and across the lipid bilayer. The proteins move about between the lipids, which are also on the move.

> **4** What do we mean by 'protein', 'lipid', 'phospholipid' and 'carbohydrate'?

Figure 2.6 Plasma membrane as a fluid mosaic structure.

Diagrammatic representation

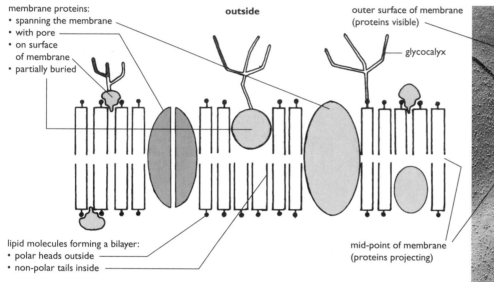

SEM of freeze-fractured membrane

Movement across the plasma membrane

Where there is a difference in concentration of atoms, molecules or ions (e.g. of a solute in a liquid), net movement will occur to areas of lower concentration. This results in the molecules becoming evenly distributed. The continuous, random movements of atoms, molecules and ions in liquids (or gases) causes this process and it is known as **diffusion**.

In **facilitated diffusion**, substances are assisted in diffusion across a membrane by the action of a particular membrane molecule, called an ionophore. Typically, an ionophore forms a pore in the membrane in the presence of a substance whose diffusion it facilitates. Examples are the facilitated diffusion of adenosine diphosphate (ADP) into mitochondria from the cytosol, and of ATP from the mitochondrial matrix into the cytosol.

Figure 2.7 Diffusion and facilitated diffusion compared.

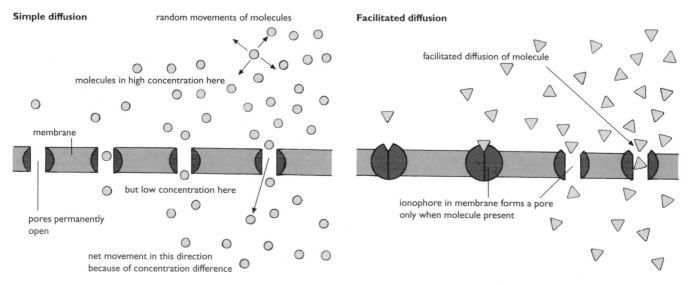

Osmosis is a special case of diffusion that involves movement of water molecules from a region of their high concentration to a region of their low concentration through a partially permeable membrane. Molecules in pure water are all free to move about, but in a solution the water molecules form a 'cloud' around solute molecules and move more slowly. The **water potential** (Ψ – Greek letter *psi*) of a solution is the tendency for water molecules to enter or leave by osmosis. Pure water has the highest value of Ψ, set arbitrarily at zero. Water moves from a region of high water potential (zero or less negative) to a region of lower water potential (more negative value). So, water enters and leaves animal (and plant) cells because the plasma membrane is partially permeable, and the internal and external solutions normally have differing water potentials.

Figure 2.8 Osmosis demonstrated.

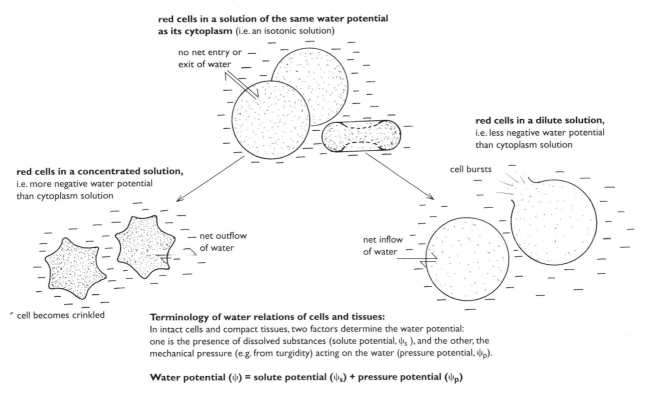

red cells in a solution of the same water potential
as its cytoplasm (i.e. an isotonic solution)

no net entry or exit of water

red cells in a dilute solution, i.e. less negative water potential than cytoplasm solution

cell bursts

red cells in a concentrated solution, i.e. more negative water potential than cytoplasm solution

net outflow of water

net inflow of water

cell becomes crinkled

Terminology of water relations of cells and tissues:
In intact cells and compact tissues, two factors determine the water potential: one is the presence of dissolved substances (solute potential, ψ_s), and the other, the mechanical pressure (e.g. from turgidity) acting on the water (pressure potential, ψ_p).

Water potential (ψ) = solute potential (ψ_s) + pressure potential (ψ_p)

By **active transport** cells can accumulate useful substances against a concentration gradient, that is, from a low concentration to a higher concentration, using energy from respiration released via ATP. Active uptake is a selective process involving protein 'pumps' in membranes. When the energy of respiration is directly involved, the pump is a primary pump. There are primary pumps specific for many ions such as Ca^{2+} and for organic compounds such as sucrose, etc.

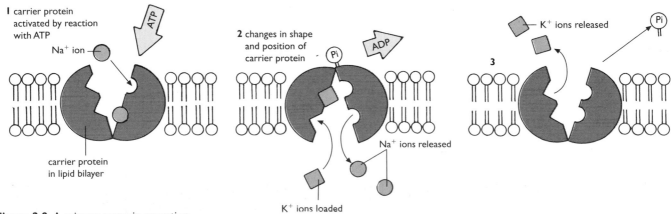

1 carrier protein activated by reaction with ATP

ATP

Na^+ ion

carrier protein in lipid bilayer

2 changes in shape and position of carrier protein

Pi

ADP

Na^+ ions released

K^+ ions loaded

K^+ ions released

Pi

3

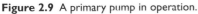

Figure 2.9 A primary pump in operation.

Bulk transport can occur by in-tucking of the membrane to form vesicles, taking in liquids by **pinocytosis** and solid particles by **phagocytosis**. This involves energy from respiration. In the mammal, phagocytosis is shown by white cells ingesting invading bacteria, and by the phagocytic cells of the liver and spleen, which remove damaged cells, such as aged red cells.

Cell specialisation

The tissues and organs of the mammal develop from the cells of the growing embryo (page 85). As groups of cells become specialised (known as **division of labour**) they develop distinctive physiological and structural differences. As a consequence, most tissues are highly specialised for particular tasks, and are dependent upon other specialised cells for survival.

Figure 2.10 The four types of tissues of mammals.

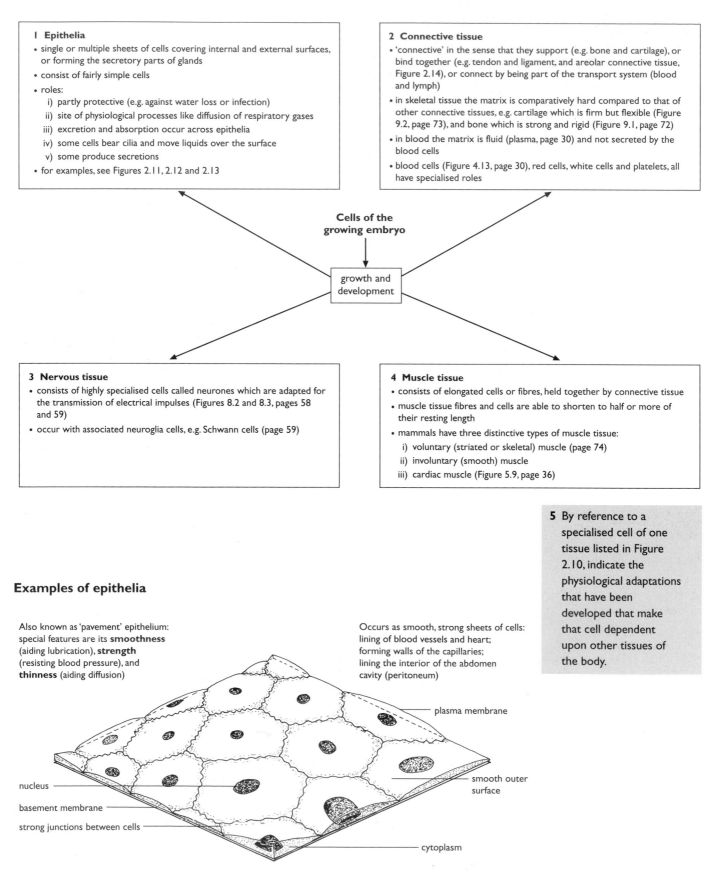

I Epithelia
- single or multiple sheets of cells covering internal and external surfaces, or forming the secretory parts of glands
- consist of fairly simple cells
- roles:
  - i) partly protective (e.g. against water loss or infection)
  - ii) site of physiological processes like diffusion of respiratory gases
  - iii) excretion and absorption occur across epithelia
  - iv) some cells bear cilia and move liquids over the surface
  - v) some produce secretions
- for examples, see Figures 2.11, 2.12 and 2.13

2 Connective tissue
- 'connective' in the sense that they support (e.g. bone and cartilage), or bind together (e.g. tendon and ligament, and areolar connective tissue, Figure 2.14), or connect by being part of the transport system (blood and lymph)
- in skeletal tissue the matrix is comparatively hard compared to that of other connective tissues, e.g. cartilage which is firm but flexible (Figure 9.2, page 73), and bone which is strong and rigid (Figure 9.1, page 72)
- in blood the matrix is fluid (plasma, page 30) and not secreted by the blood cells
- blood cells (Figure 4.13, page 30), red cells, white cells and platelets, all have specialised roles

Cells of the growing embryo

growth and development

3 Nervous tissue
- consists of highly specialised cells called neurones which are adapted for the transmission of electrical impulses (Figures 8.2 and 8.3, pages 58 and 59)
- occur with associated neuroglia cells, e.g. Schwann cells (page 59)

4 Muscle tissue
- consists of elongated cells or fibres, held together by connective tissue
- muscle tissue fibres and cells are able to shorten to half or more of their resting length
- mammals have three distinctive types of muscle tissue:
  - i) voluntary (striated or skeletal) muscle (page 74)
  - ii) involuntary (smooth) muscle
  - iii) cardiac muscle (Figure 5.9, page 36)

5 By reference to a specialised cell of one tissue listed in Figure 2.10, indicate the physiological adaptations that have been developed that make that cell dependent upon other tissues of the body.

Examples of epithelia

Also known as 'pavement' epithelium: special features are its **smoothness** (aiding lubrication), **strength** (resisting blood pressure), and **thinness** (aiding diffusion)

Occurs as smooth, strong sheets of cells: lining of blood vessels and heart; forming walls of the capillaries; lining the interior of the abdomen cavity (peritoneum)

plasma membrane

nucleus

smooth outer surface

basement membrane

strong junctions between cells

cytoplasm

Figure 2.11 Squamous epithelium.

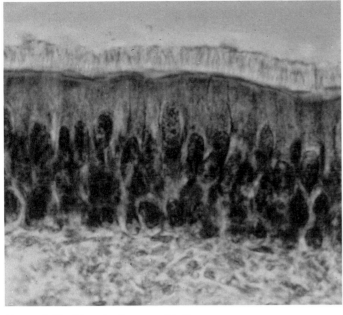

Figure 2.12 Ciliated columnar epithelium.

Ciliated epithelia line many respiratory surfaces. They often contain mucus-secreting goblet cells, and the cilia move a stream of mucus

cilia

columnar epithelium cell

nucleus

cytoplasm

This type of ciliated columnar epithelium occurs in the trachea and is called **'pseudostratified'** for here the columnar cells are of different lengths, and their nuclei occur at different levels

basement membrane

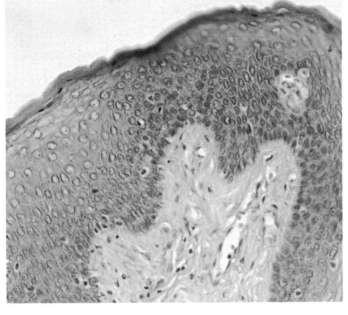

Figure 2.13 Stratified epithelium.

Stratified epithelium forms the tough impervious barrier of the external skin and the lining to mouth, anus and vagina. Cells of the outermost layer are dead, and are steadily replaced by cells from the generative layer. Proteins of the cells are turned to keratin (waterproof and tough)

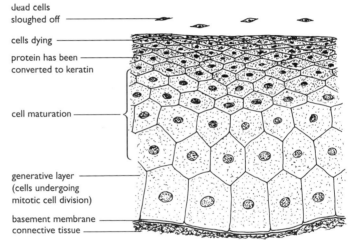

dead cells sloughed off

cells dying

protein has been converted to keratin

cell maturation

generative layer (cells undergoing mitotic cell division)

basement membrane

connective tissue

Areolar connective tissue

The role of this tissue is to bind tissues and organs together.

It consists of a matrix containing flexible fibres, some elastic and some non-elastic, and cells.

The cells lay down and maintain the tissue, but they also help protect against disease and they assist in the body's response to mechanical injury.

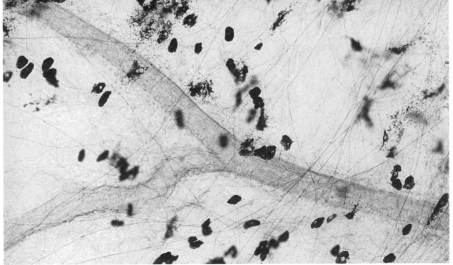

semi-fluid matrix containing:
- bundles of white collagen fibres
- a network of yellow elastic fibres

cells present:
- fibre-synthesising cells
- mast cells (secrete blood anticoagulant)
- fat-store cells
- macrophages (amoeboid cells that engulf bacteria)

Figure 2.14 Areolar connective tissue.

3 Nutrition

A mammal takes in food, which is complex organic matter, and digests it in the gut, producing molecules suitable for absorption into the blood or lymph. This is known as **holozoic nutrition** (meaning 'feeding like an animal'). Holozoic nutrition is one form of **heterotrophic nutrition**, meaning 'feeding on complex, ready-made foods' to obtain the required nutrients.

The important ecological point to remember about all heterotrophs is their dependence, directly or indirectly, on the **autotrophs**, which are organisms that manufacture their own 'elaborated' foods (mostly the photosynthetic green plants).

1 Can you think of two other forms of heterotrophic nutrition?

2 In what ways are all heterotrophs dependent on autotrophs?

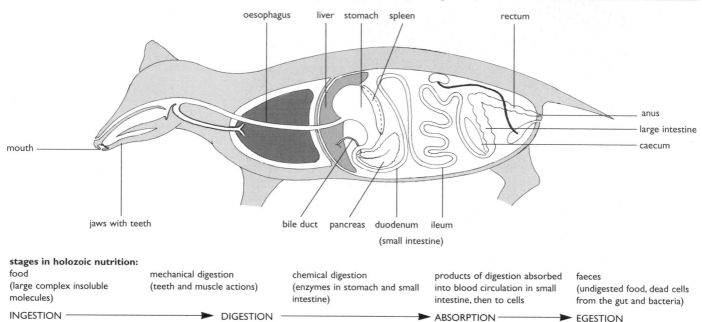

stages in holozoic nutrition:

| food (large complex insoluble molecules) | mechanical digestion (teeth and muscle actions) | chemical digestion (enzymes in stomach and small intestine) | products of digestion absorbed into blood circulation in small intestine, then to cells | faeces (undigested food, dead cells from the gut and bacteria) |

INGESTION ⟶ DIGESTION ⟶ ABSORPTION ⟶ EGESTION
⟶ ASSIMILATION

Roles of the gut in nutrition

The gut is a long, hollow muscular tube connecting mouth to anus. Each region of the gut is specialised for particular steps in nutrition, but the gut as a whole is the site of:

- movement of partly digested food through the body by waves of muscular contractions (**peristalsis**, Figure 3.4);
- digestion by **enzymes** added to the broken-up food, causing chemical breakdown;
- absorption of the products of digestion into the body, mostly in the ileum (page 20);
- a huge 'flora' of microorganisms, present for most of the length of the gut. Some microorganisms have roles in metabolism of the host, e.g. vitamin K production in the colon, and cellulose digestion in the stomachs of cows (page 23). The most prevalent bacterium in the human gut is a harmless strain of *Escherichia coli* (referred to as *E. coli*).

The **layered structure of the gut wall** is shown in Figure 3.2. The ways that this basic structure is modified in different regions are shown on the following pages, along with the glands associated with digestion.

Figure 3.1 Rat in saggital section, showing layout of gut and the stages involved in holozoic nutrition.

Figure 3.2 Gut wall, dissected into its component layers.

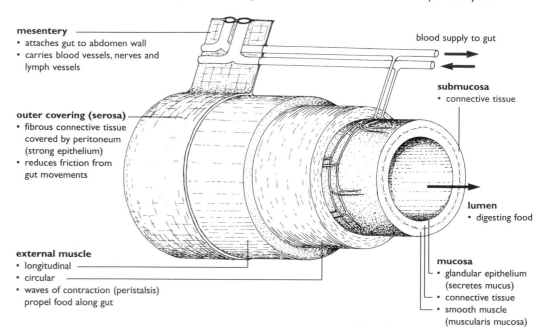

Oesophagus

This is a straight, thick-walled, muscular region of the gut connecting the pharynx (rear of mouth cavity) with the stomach. It has a glandular epithelium that lubricates the passage of food by secretion of copious quantities of mucus. The waves of peristalsis passing down the oesophagus are triggered by a ball of food matter (a bolus) entering from the pharynx.

Figure 3.3 Structure of the human oesophagus.

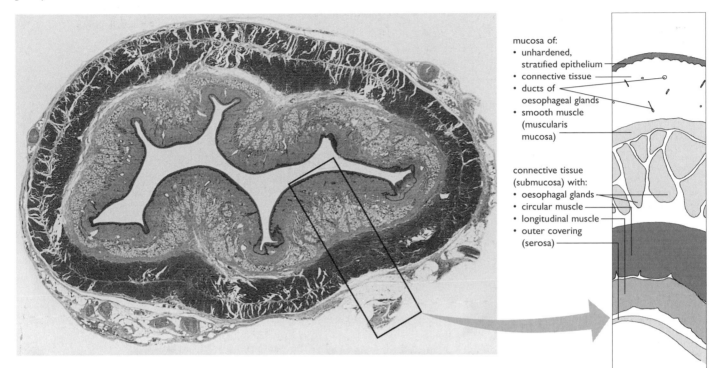

mucosa of:
- unhardened, stratified epithelium
- connective tissue
- ducts of oesophageal glands
- smooth muscle (muscularis mucosa)

connective tissue (submucosa) with:
- oesophagal glands
- circular muscle
- longitudinal muscle
- outer covering (serosa)

1 Movement of food from mouth to oesophagus

- lips, tongue and teeth work together to move food about the mouth and to break up the food
- the broken up lumps of food are moistened and bound together by saliva from the salivary glands
- the tongue presses the food against the hard palate, forming it into a lubricated 'ball', called a bolus
- swallowing of food is triggered by the voluntary action of moving the bolus to the back of the mouth
- once the food is in the oesophagus it is moved by reflex action

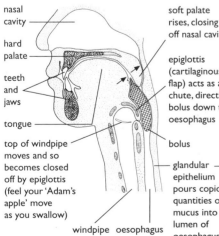

nasal cavity

hard palate

teeth and jaws

tongue

top of windpipe moves and so becomes closed off by epiglottis (feel your 'Adam's apple' move as you swallow)

windpipe   oesophagus

soft palate rises, closing off nasal cavity

epiglottis (cartilaginous flap) acts as a chute, directing bolus down the oesophagus

bolus

glandular epithelium pours copious quantities of mucus into lumen of oesophagus

2 Movement of bolus down the oesophagus by peristalsis (wave of muscular contraction and relaxation)

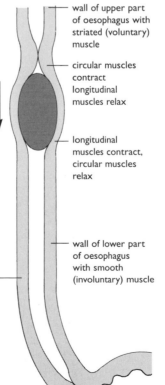

wall of upper part of oesophagus with striated (voluntary) muscle

circular muscles contract longitudinal muscles relax

longitudinal muscles contract, circular muscles relax

wall of lower part of oesophagus with smooth (involuntary) muscle

3 Entry of bolus into stomach

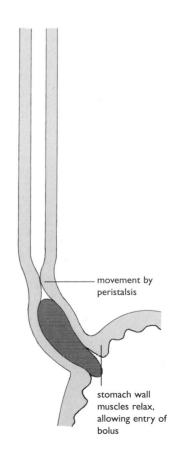

movement by peristalsis

stomach wall muscles relax, allowing entry of bolus

Figure 3.4 Movement of food from mouth to stomach in humans.

Digestion in the mouth

Digestion begins with the mechanical break-up of food by the actions of teeth in the mouth (buccal cavity). This process is known as **mastication**. You can get a good idea of the ways teeth of different mammals work by examining preserved skulls (or realistic plastic models). The position of the teeth, the shapes of cutting and grinding surfaces, and the plane of movement of the lower jaw are all important.

3 What structural feature in mammals permits the retention of food in the mouth, and allows mastication by specialised teeth?

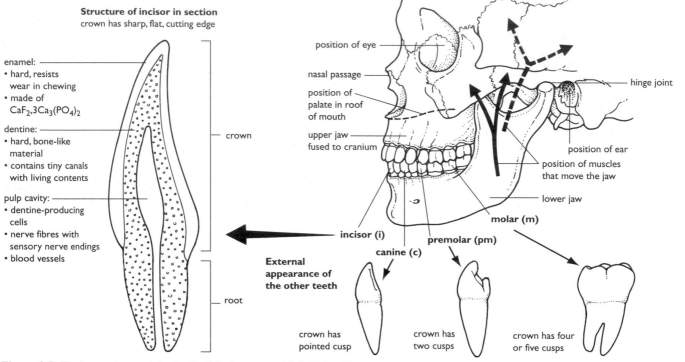

Figure 3.5 The human jaws and the teeth of the 'permanent' (adult) dentition.

Most mammals have **two sets of teeth** in their lifetime, a deciduous set followed by a permanent set. The numbers of teeth of each type in a dentition in one side of the jaw are represented in a dental formula. The dental formulae of humans are:

4 What is the advantage to later digestion of food being broken into tiny pieces at the outset?

$$\text{Deciduous set:} \quad i\,\frac{2}{2} \; c\,\frac{1}{1} \; m\,\frac{2}{2} \qquad \text{Permanent set:} \quad i\,\frac{2}{2} \; c\,\frac{1}{1} \; pm\,\frac{2}{2} \; m\,\frac{3}{3}$$

('milk set')

The photograph of an X-ray plate below shows the dentition of a 10-year-old girl whose permanent dentition was well advanced. Replacement starts at around 5–6 years, but takes several years to complete.

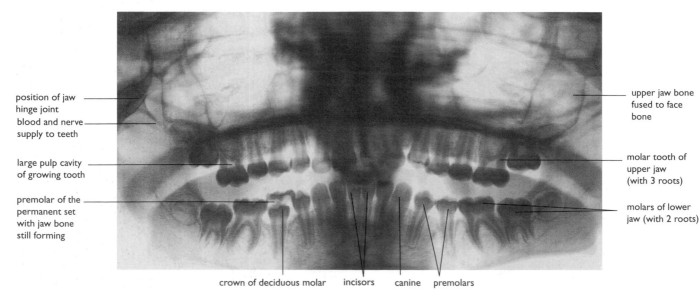

Figure 3.6 Human dentition in transition (deciduous to permanent set).

The skulls of herbivores and carnivores

Both the position of the eyes and the type of dentition shown by these skulls give clues about the feeding processes of these animals.

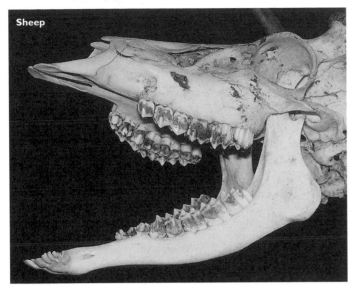

Sheep

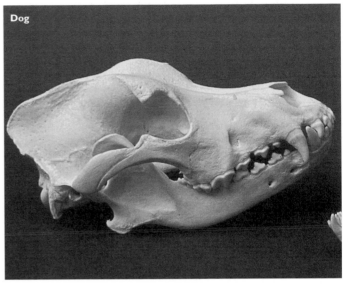

Dog

Dentition of sheep:

$$i = \frac{0}{3} \quad c = \frac{0}{1} \quad pm = \frac{3}{3} \quad m = \frac{3}{3}$$

- incisors and canine of lower jaw bite against horny pad in the upper jaw
- premolars and molars form a grinding mill
- the surface of the teeth wear to form sharp ridges
- teeth continue to grow throughout life
- during chewing the jaw moves sideways

Dentition of dog:

$$i = \frac{3}{3} \quad c = \frac{1}{1} \quad pm = \frac{4}{4} \quad m = \frac{2}{3}$$

- incisors are sharp, and used to nip, grip and tear
- canine teeth are curved and sharply pointed; used to seize and kill prey and tear flesh
- premolars and molars are used to cut and crush
- some premolars and molars form a shearing unit (like scissors blades); called the carnassial teeth

Figure 3.7 The skulls of a herbivore and a carnivore.

Saliva from salivary glands

Salivary glands are glands with ducts (exocrine glands). Saliva is a watery solution containing salts, mucus, the enzyme amylase and the antibacterial enzyme lysozyme. This secretion allows the mouth (tongue, lips and teeth) and throat to function in speech and in the movement of food. Lysozyme reduces bacterial contamination.

Salivary glands are positioned in the walls of the mouth, under the tongue and in the neck. They are a type of gut gland situated outside the gut wall (as is the pancreas, page 19). Approximately 1.5 dm^3 of saliva is secreted each day by a human.

5 What substance does amylase begin the digestion of?

Figure 3.8 The structure of the salivary glands.

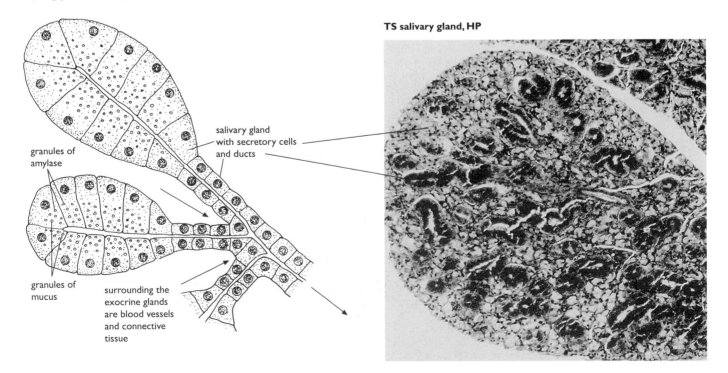

granules of amylase

salivary gland with secretory cells and ducts

granules of mucus

surrounding the exocrine glands are blood vessels and connective tissue

TS salivary gland, HP

Stomach

The stomach is a muscular bag that can expand to contain a meal. The inner lining of the stomach wall is rich in mucus-secreting cells, and has thousands of tiny pits called **gastric glands**. The chief roles of the stomach are to store food temporarily, and to mix it with gastric juice. Mechanical digestion occurs by the churning action of waves of peristalsis passing across the gastric wall. Enzymic **digestion of proteins** begins in the stomach. It is an extremely acid environment because of secretions from the gastric glands. These secretions are coordinated with the presence of food (page 22). A protein-rich meal typically remains in the stomach for about 4 hours; in this time the stomach contents become a semi-liquid known as **chyme**.

6 State two foods of
   a) plant origin,
   b) animal origin
   that are rich in proteins and will therefore start being digested in the stomach.

7 What is the chief role of protein in the diet?

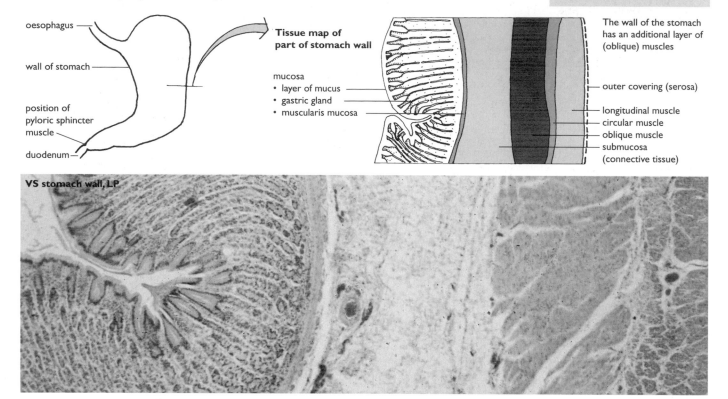

Figure 3.9 Structure of the stomach wall.

A layer of **mucus** forms the only barrier between the tissues of the stomach wall and the food taken into the stomach. The stomach wall becomes bathed in corrosive, digestive gastric juice in the strongly acidic environment, but the mucus prevents its self-digestion (autolysis). Very rarely the mucus lining may be breached, possibly leading to a gastric ulcer. A bacterium, *Helicobacter pylori*, is involved in the processes of ulcer formation.

Despite the protection provided by mucus, the cells of the lining of the stomach (epithelium and gastric glands) last for only a few days and are constantly replaced.

Figure 3.10 Gastric epithelium.

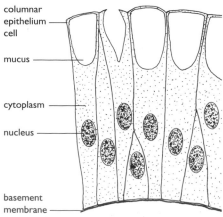

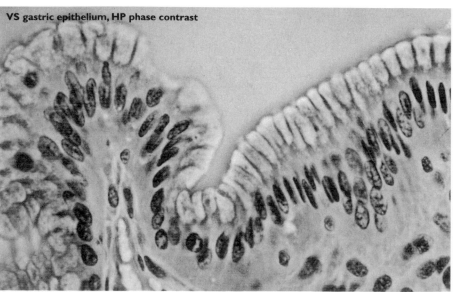

Gastric glands

The stomach lining contains millions of pits called gastric glands. Typically, more than 0.5 dm³ of gastric juice is secreted per meal, produced by this huge battery of tiny glands. The different types of cell in the gastric glands secrete a particular component of gastric juice, each having a specific role in the digestion processes of the stomach.

Figure 3.11 Gastric glands and the formation of gastric juice.

TS gastric gland, HP

Gastric gland in LS

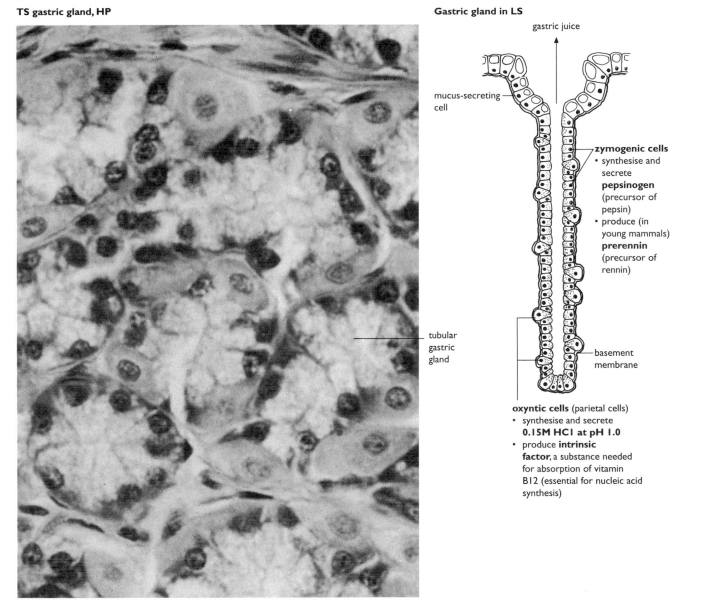

gastric juice

mucus-secreting cell

zymogenic cells
• synthesise and secrete **pepsinogen** (precursor of pepsin)
• produce (in young mammals) **prerennin** (precursor of rennin)

tubular gastric gland

basement membrane

oxyntic cells (parietal cells)
• synthesise and secrete **0.15M HCl at pH 1.0**
• produce **intrinsic factor**, a substance needed for absorption of vitamin B12 (essential for nucleic acid synthesis)

Table 3.1 Roles of the active components of gastric juice.

Component	Action
Hydrochloric acid	establishes the optimum pH for the enzymes of the stomach (pH 1.5–2.0 in chyme); kills bacteria, so that the stomach is largely bacteriologically sterile; helps denature protein and softens fibrous connective tissue in foods; activates pepsin; activates rennin; converts calcium and iron salts into forms suitable for absorption; commences the hydrolysis of sucrose; splits nucleoproteins into component proteins and nucleic acids
Pepsin	hydrolyses proteins in the diet to shorter chain polypeptides. The breaks in the protein (↑) occur mostly beside tyrosine (Tyr) and phenylalanine (Phe) residues: (Leu)–(Val)–(Glu)–(Ser)–(Gly)–(Val)–(Tyr)–(Ala)–(Ala)–(Phe)–(Gly)–(Glu)– amino acid residues peptide linkages This is the action of an endopeptidase (i.e. selective hydrolysis of peptide linkages *within* the protein macromolecules)
Rennin (young mammals)	commences the digestion of milk protein (caseinogen) by coagulating it. In the coagulated form proteins are available for hydrolysis, starting with the action of pepsin

Duodenum: the first part of the small intestine

From the stomach the chyme passes into the duodenum where the bulk of digestion is completed. The human small intestine is about 5 m long, of which the duodenum takes up only 30 cm. The chyme is squirted in, a little at a time. Enzymes from the pancreas, via the pancreatic duct, and bile from the liver, via the bile duct, are added at this stage.

In the wall of the duodenum the submucosa is thrown into quite large inwardly projecting ridges. These ridges support a vast number of **villi**, which project into the lumen. At the base of the villi, located in the mucosa, are simple tubular glands called the **crypts of Lieberkühn**. In the sub-mucosa of the duodenum there are small, rounded **Brunner's glands**. Goblet cells in both these types of gland secrete **mucus and alkaline salts**. This secretion briefly protects the lining of the intestine, both mechanically and from digestion by stomach enzymes.

> **8** What is the practical effect of the presence of villi in the intestine?

Figure 3.12 Structure of the duodenum.

VS wall of duodenum, LP

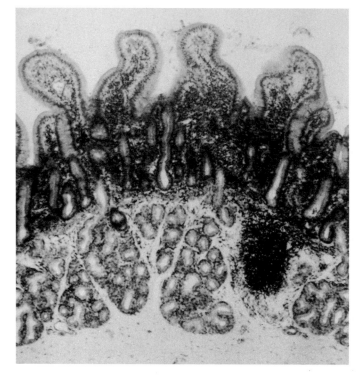

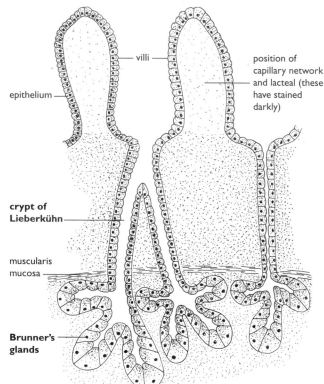

The cells of the epithelium of the small intestine survive only for about 2 days *in situ*. Once damaged, they are sloughed off into the gut lumen, and replaced by cells from the crypts of Lieberkühn that migrate up to the villi.

Figure 3.13 VS villus, HP, showing epithelium with goblet cells.

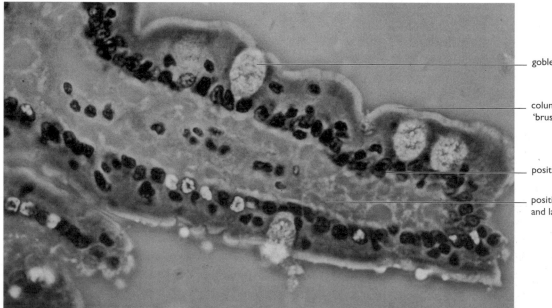

Bile is produced in the liver (page 52) and collects in the bile ducts (and gall bladder in humans and some other mammals). Bile enters the duodenum when acid chyme arrives from the stomach. Bile is a yellow–green fluid consisting of bile salts, bile pigments, cholesterol, salts and mucus (no digestive enzymes). Bile acts to:

- neutralise the acidity of chyme by the presence of alkaline salts;
- create tiny droplets of liquid (emulsification) by lowering the surface tension of lipid globules.

The result is that lipids in the chyme are at the correct pH, and present a huge surface area for hydrolysis by lipase.

The **pancreas** is largely an exocrine gland secreting **pancreatic juice** via a duct into the duodenum. Pancreatic juice is an alkaline solution of digestive enzymes (pH 7.5–8.8). Its buffer action is due to the presence of hydrogencarbonate and other alkaline ions in aqueous solution, the main neutralising agents.

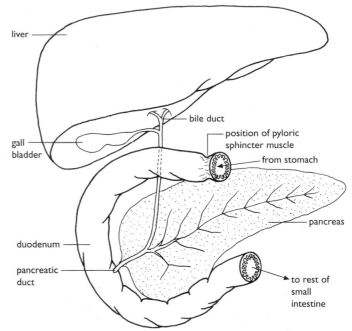

Figure 3.14 How bile and pancreatic juice reach the duodenum.

Table 3.2 The role of the hydrolytic enzymes of pancreatic juice.

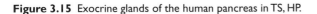

Enzyme	Substrate	Product(s)
Amylase	starch	maltose
*Trypsin	proteins	polypeptides
†Chymotrypsin	proteins	polypeptides
Peptidases	polypeptides	amino acids
Lipase	lipids	fatty acids and glycerol
Nucleases	nucleic acid	nucleotides

*Trypsinogen is activated by enteropeptidase, an enzyme secreted by the epithelium of the small intestine, to produce trypsin; †chymotrypsinogen is activated by trypsin in the small intestine to produce chymotrypsin.

Figure 3.15 Exocrine glands of the human pancreas in TS, HP.

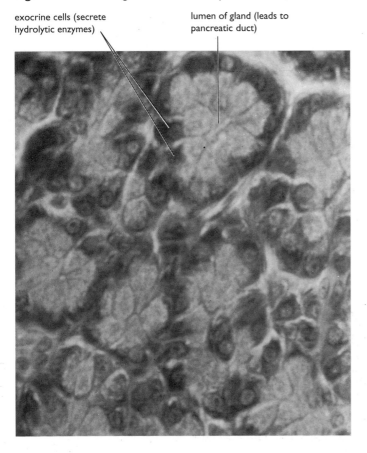

exocrine cells (secrete hydrolytic enzymes)

lumen of gland (leads to pancreatic duct)

Figure 3.16 A summary of digestion and absorption of lipid.

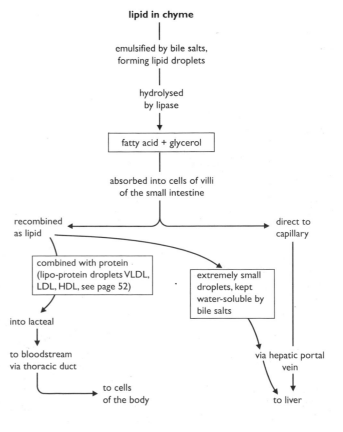

Ileum: absorption in the small intestine

In the remainder of the small intestine, known as the ileum, the ridges (seen in the duodenum) disappear. The inner surface of the ileum consists of a 'forest' of **villi**. The villi are supplied with capillaries and lymph vessels (part of the lymphatic system). Smooth muscle fibres, extending into the villi from the muscularis mucosa, contract and relax, and move the villi among the digested food. The outer surface of the epithelial cells lining the villi have a 'brush border' of 2000 or so **microvilli**, which increase the surface area for absorption.

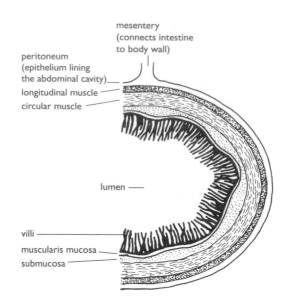

Figure 3.17 The structure of the ileum.

LS ileum wall, LP

Preparation showing capillaries, HP. Capillary networks have been injected with coloured latex. *Note:* the lacteals are not visible in this preparation

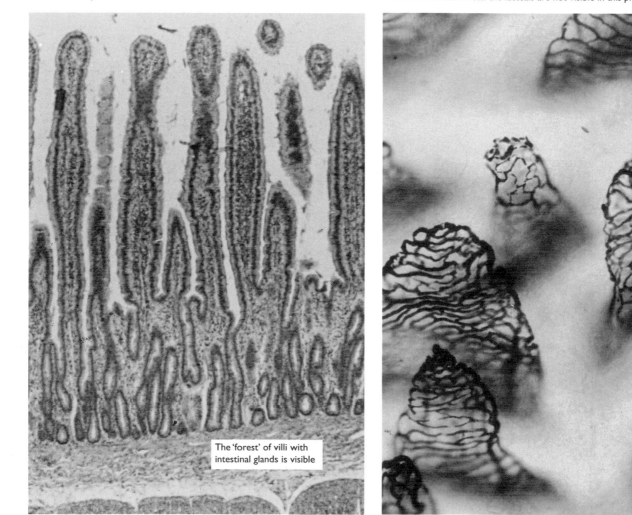

The 'forest' of villi with intestinal glands is visible

Figure 3.18 The structure of villi.

Digestion in the ileum is completed by enzymes from the following sources:

- enzymes of the pancreatic juice, introduced in the duodenum, and still active in the ileum;
- enzymes bound to the membrane of the microvilli of the epithelial lining (see opposite);
- enzymes bound to the membrane of the microvilli released from the dislodged and disintegrating epithelial cells, functioning in the lumen.

9 Where does lymph in the lymph vessels of the small intestine drain back into the blood circulation?

Figure 3.19 The detailed structure of a villus.

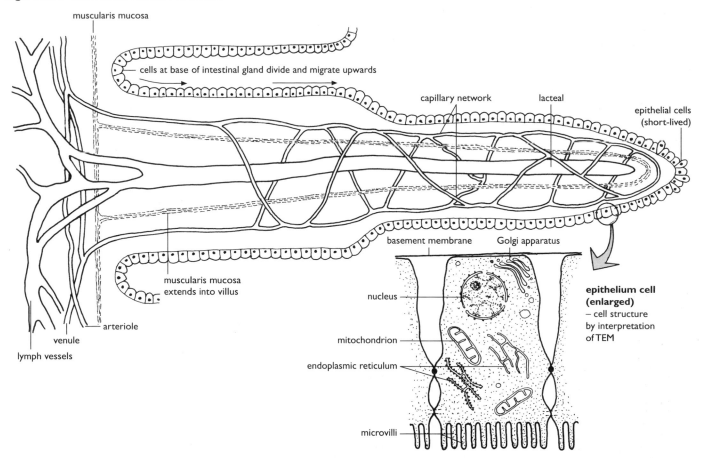

Absorption into the epithelial cells of the villi is the final step in the digestion of proteins and carbohydrates. It is catalysed by special proteins in the plasma membrane of the microvilli, which function as both **hydrolytic enzymes and carrier proteins**. The products of lipid digestion are largely absorbed into the lacteal (see Figure 3.16).

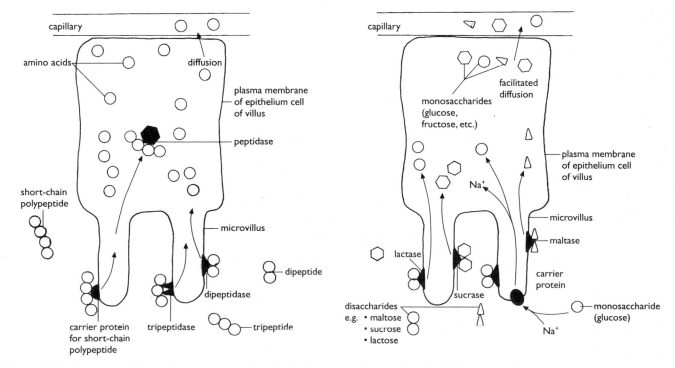

Figure 3.20 Absorption of the products of protein digestion (left) and carbohydrate digestion (right).

In the **colon** water and mineral salts are absorbed. What remains, undigested food (plant fibre is a major component) together with sloughed off cells of the gut epithelium and the microflora of the gut, is now known as **faeces**.

Control of feeding and digestion

The hypothalamus (situated at the base of the forebrain) is where regulation of internal body activities and feeding occurs. Levels of metabolites in the blood are monitored here. The 'feeding centre' of the hypothalamus is potentially constantly active, but a neighbouring 'satiety centre' becomes active after a meal and suppresses activity of the 'feeding centre' for a time. Impulses from the 'feeding centre' run to a regulatory centre in the medulla oblongata (of the hindbrain). From the medulla, the **vagus nerve** runs to the internal organs, such as the gut, with nerve fibres of the autonomic nervous system (sympathetic and parasympathetic nervous systems). Activity in sympathetic nerves to the gut **decreases** glandular secretions. Activity in parasympathetic nerves to the gut **increases** muscular activity (peristalsis, etc.) and glandular secretions. The secretion of digestive enzymes by the stomach and the pancreas is also regulated by specific hormones.

10 What do we mean by an 'autonomic nervous system'? What is the alternative nervous system?

Figure 3.21 Case studies of the regulation of secretion of digestive juices.

Pancreatic juice and bile secretion

Chyme rich in amino acids and fats stimulates cells of the duodenal mucosa to secrete the hormone 'cholecystokinin' (pancreozymin) into the bloodstream, where it is carried around the body. At the pancreas (exocrine cells), this hormone stimulates secretion of juice rich in pancreatic enzymes. At the gall bladder/bile duct, this hormone stimulates the secretion of bile

Acid chyme entering the duodenum stimulates cells of the duodenal mucosa to secrete the hormone 'secretin'. At the pancreas this hormone stimulates secretion of juice rich in Na_2HCO_3 (strongly alkaline secretion)

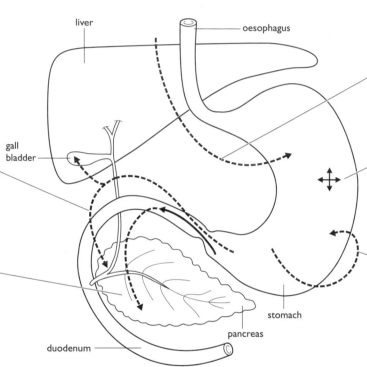

Gastric juice secretion

The sight of food and/or the presence of food in the mouth triggers nervous stimulation of the gastric glands which secrete gastric juice (parasympathetic fibres of the vagus nerve)

Arrival of food in the stomach stretches the stomach wall (mechanical stimulation) and triggers secretion of gastric juice

A hormone 'gastrin' is secreted by cells of the gastric epithelium when food is in the stomach. Gastrin enters the bloodstream, circulates to the gastric glands, and triggers secretion of gastric juice

Digestion of cellulose

Cellulose is the most abundant organic compound in the biosphere; it makes up more than 50% of all the 'organic' carbon (CO_2 is 'inorganic' carbon). It forms the bulk of green plant cell walls, and also exists in leaf-litter and much of the decaying organic matter in the soil. Mammals do not have cellulase enzymes and so herbivorous and omnivorous mammals cannot themselves digest plant cell walls. However, many bacteria (and many fungi and protozoa) do produce cellulase. Herbivorous mammals exploit this facility of bacteria to digest cellulose.

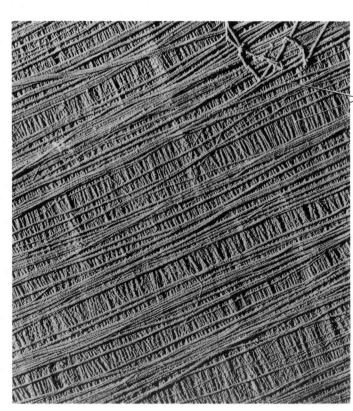

Figure 3.22 TEM of cellulose (× 30 000).

microfibrils (many layers)

11 Why is the relationship between gut bacteria and herbivorous mammals described as 'mutualistic'?

Ruminants

Ruminants (e.g. cows and sheep) have a four-chambered 'stomach'. The first three chambers (rumen–reticulum–omasum) are derived from the lower part of the oesophagus. The final chamber (abomasum) is the true stomach.

Figure 3.23 Digestion of cellulose by ruminants.

② the **rumen** functions as a 'fermentation vat' under anaerobic conditions. A huge flora of bacteria produce cellulase and other digestive enzymes, and convert the cellulose present to sugars. These sugars are fermented to organic acids which are then absorbed into the blood through the rumen wall. They are the energy source for the ruminant. Waste products: CO_2, water vapour and methane (all are 'greenhouse' gases) are belched out.
The bacteria also produce proteins from ammonium salts, and vitamins are synthesised.
The rumen also houses protozoa known as ciliates that feed on the bacterial 'flora' present

③ the fermented grass is passed to the **reticulum**. Here it is formed into balls known as 'cud'. The cud is regurgitated to the mouth for further grinding up by the teeth before being swallowed again

① grass cropped, ground up by premolars and molars, and mixed with saliva in the **mouth**. Then swallowed, and passed to the rumen

④ in the **omasum** water from the cud is reabsorbed (typically a cow secretes approximately 150 litres of saliva per day). The more solidified food material is passed to the abomasum

⑤ in the **abomasum** normal gastric secretions begin digestion of the proteins from plant cells, bacteria and ciliates

⑥ the chyme now passes to the **duodenum** where the bulk of digestion is completed

Non-ruminants

In non-ruminants (e.g. rabbits and horses) the cellulose-digesting bacteria are housed in the caecum and appendix. During the journey through the gut, the digesting food is diverted towards the caecum; in by peristalsis, out by reverse peristalsis. Here it meets the bacterial 'flora' in an anaerobic environment. Cellulose is converted into glucose and then into organic acids.

In the horse the products of digestion are absorbed in the colon.

In rabbits, faecal pellets egested overnight are eaten as they emerge from the anus. They are re-digested, and nutrients released by bacterial action on the chewed and partly digested plant cells are absorbed in the small intestine. When the waste is finally eliminated it forms hard, dry faecal pellets typically seen on grassland where rabbits feed.

Figure 3.24 TS of the appendix of a rabbit, LP.

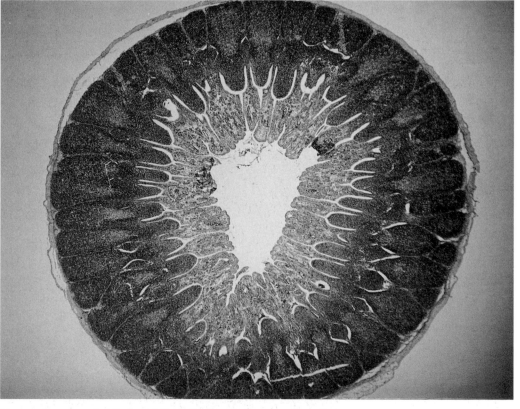

In young humans extensive lymph tissue develops in the appendix wall

Exchange and transport

Energy is released in cells by the oxidation of sugars and other substances, in a process called **respiration**. Respiration occurs in every living cell, and provides the chemical energy source (ATP) that drives the cellular machinery. As a consequence, gases are exchanged between respiring cells and the environment.

Mammals have an external body surface through which any **gaseous exchange** is impossible. They are relatively large, compact animals with a high rate of metabolism; this means that they require a lot of oxygen and produce a lot of carbon dioxide per unit of body volume. Gaseous exchange in mammals takes place in special organs and an efficient system involving:

1 **lungs**, compact internal organs in which the surface for gaseous exchange is greatly increased;
2 a **ventilation mechanism** (breathing), which moves a fresh supply of air to the respiratory surface;
3 an **internal transport system** (blood circulation), moving respiratory gases between lungs and respiring cells;
4 a **respiratory pigment**, increasing the oxygen-carrying capacity of the blood.

The **lungs of mammals** are enclosed in an air-tight body compartment called the thorax, with sides made of the ribs and a 'floor' made of the dome-shaped diaphragm. The ribs are moved upwards and outwards, and the diaphragm flattens by contraction. These movements increase the volume of the thorax (and therefore lower the air pressure). Relaxation of these muscles decreases the volume of the thorax (and increases the air pressure). The inner surface of the thorax and the outer surface of the lungs are covered by a smooth pleural membrane, and the space between contains lubricating pleural fluid.

The lungs are a pair of delicate, elastic organs of millions of tiny air sacs (alveoli) served by tiny tubes (alveolar ducts and bronchioles) connected to the atmosphere by the main air supply tubes (the bronchi and trachea). Air enters the lungs (we breathe in) when the internal pressure is lower than the external pressure. Air leaves (we breathe out) when the internal pressure is higher than the external pressure.

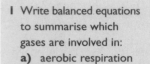

1 Write balanced equations to summarise which gases are involved in:
 a) aerobic respiration
 b) anaerobic respiration by lactic acid fermentation.

Figure 4.1 The thorax, and the mechanism of lung ventilation in humans.

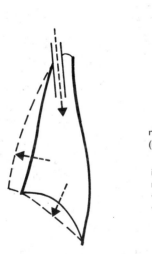

inspiration:
• external intercostal muscles contract
• internal intercostal muscles relax
• diaphragm muscles contract

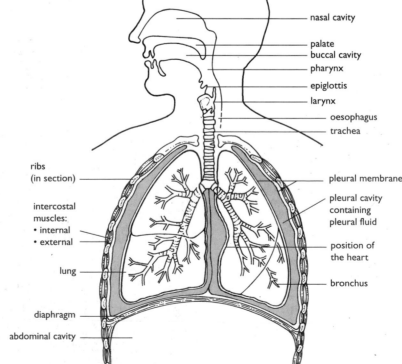

nasal cavity
palate
buccal cavity
pharynx
epiglottis
larynx
oesophagus
trachea
pleural membrane
pleural cavity containing pleural fluid
position of the heart
bronchus

ribs (in section)
intercostal muscles:
• internal
• external
lung
diaphragm
abdominal cavity

expiration:
• external intercostal muscles relax
• internal intercostal muscles contract
• diaphragm muscles relax

volume of thorax (and therefore of lungs) increases; pressure is reduced below that of atmospheric pressure and air flows in

volume of the thorax (and therefore of lungs) is decreased; pressure is increased above that of atmospheric pressure and air flows out

Breathing

Ventilation of the thorax is controlled by the respiratory centre in the medulla of the hindbrain. Here, adjacent inspiration and expiration centres interact, and from here branches of cranial nerves run to the ribs and diaphragm muscles. Breathing occurs automatically by involuntary reflex actions, but the rate is continually adjusted.

Table 4.1 Percentage composition of inspired, alveolar and expired air.

Gas	Inspired	Alveolar	Expired
O_2	20.95	13.80	16.40
CO_2	0.04	5.50	4.00
N_2	79.01	80.70	79.60
H_2O	Variable	Saturated	Saturated

A recording **spirometer** is used to analyse the pattern of change in lung volume during human breathing by measuring parameters such as inspiratory capacity and tidal volume. The movements of the lid of the air-tight spirometer chamber are recorded via a position transducer, connecting box and microcomputer. The results are printed out using appropriate control software.

Figure 4.3 Investigation of breathing.

Figure 4.2 The regulation of breathing.

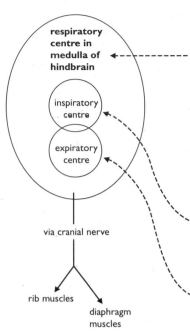

breathing occurs automatically by involuntary reflex action, but can be over-ridden by messages from the **cerebral hemispheres**, e.g. when talking, singing, playing the saxophone, etc.

breathing rate is continually adjusted to meet the body's changing needs, e.g. different levels of physical activity:

1 when breathing too slowly or when very physically active, **chemoreceptors** in medulla, carotid artery (carotid bodies), and aorta (aortic bodies) detect rising CO_2 concentration. Impulses trigger faster/deeper breathing

2 when breathing too heavily, **stretch receptors** in lungs detect excessive stretching. Impulses suppress inspiration

Human lung capacity

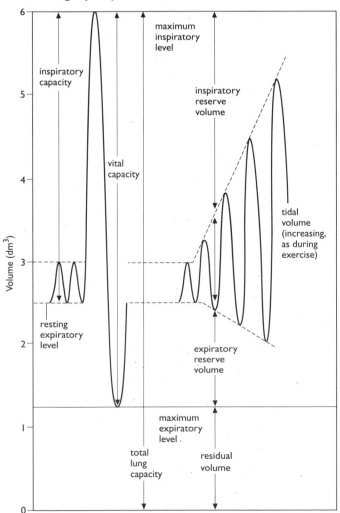

Using a recording spirometer

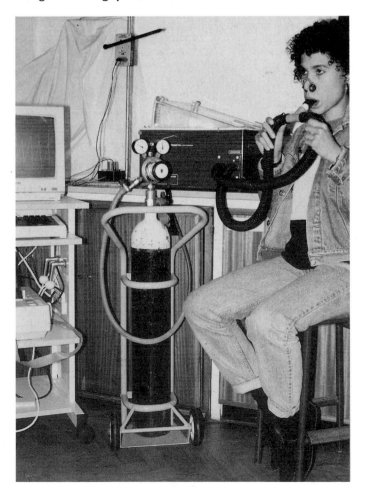

Trachea

The trachea is the start of a system of tubes of decreasing diameter, ending in clusters of tiny air sacs that make up the lung tissue. C-rings of cartilage arranged horizontally in the trachea wall prevent closure of the trachea (which may occur when food passes down the oesophagus, or if there is a sudden air pressure change during inspiration), and 'ballooning' during expiration.

The wall of the trachea also contains muscle and elastic fibres that regulate its diameter. The lining is columnar epithelium with goblet cells; mucous glands secrete onto the epithelium.

2 What are the likely roles of the ciliated epithelium and the mucus supply to the lining of the trachea?

Figure 4.4 The bronchial tree.

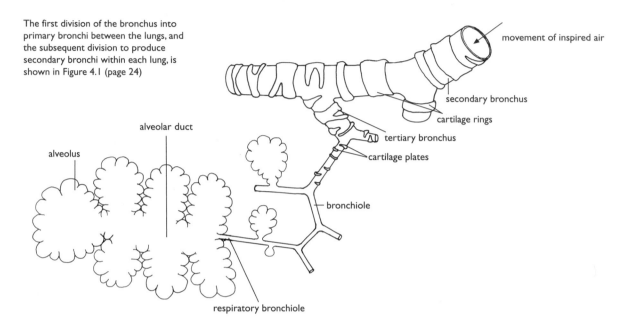

The first division of the bronchus into primary bronchi between the lungs, and the subsequent division to produce secondary bronchi within each lung, is shown in Figure 4.1 (page 24)

alveolus
alveolar duct
movement of inspired air
secondary bronchus
cartilage rings
tertiary bronchus
cartilage plates
bronchiole
respiratory bronchiole

Figure 4.5 Structure of the trachea in TS, LP.

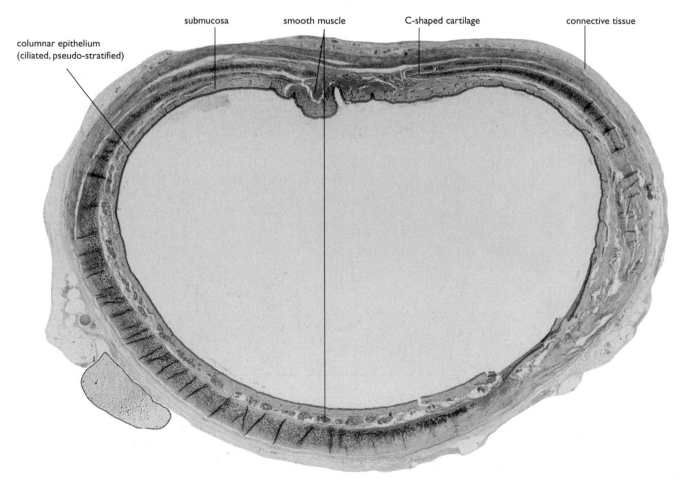

columnar epithelium (ciliated, pseudo-stratified)
submucosa
smooth muscle
C-shaped cartilage
connective tissue

Figure 4.6 Detail of the ciliated epithelium with goblet cells, HP.

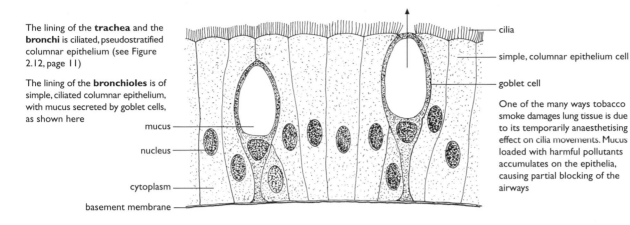

The lining of the **trachea** and the **bronchi** is ciliated, pseudostratified columnar epithelium (see Figure 2.12, page 11)

The lining of the **bronchioles** is of simple, ciliated columnar epithelium, with mucus secreted by goblet cells, as shown here

mucus

nucleus

cytoplasm

basement membrane

cilia

simple, columnar epithelium cell

goblet cell

One of the many ways tobacco smoke damages lung tissue is due to its temporarily anaesthetising effect on cilia movements. Mucus loaded with harmful pollutants accumulates on the epithelia, causing partial blocking of the airways

Lungs

Within the lungs, the secondary and tertiary bronchi are supported against collapse by fine plates of cartilage, but in the bronchioles only the smooth muscle and elastic tissues prevent collapse. The bronchioles further divide, leading to fine respiratory bronchioles, which in turn subdivide into several alveolar ducts. The alveolar ducts give rise to numerous air sacs, the alveoli, with elastic walls. The human lungs have a total surface area of about 100–150 m^2. The capillary network to the alveoli is illustrated on the next page.

Figure 4.7 TS of a fetal lung, LP.

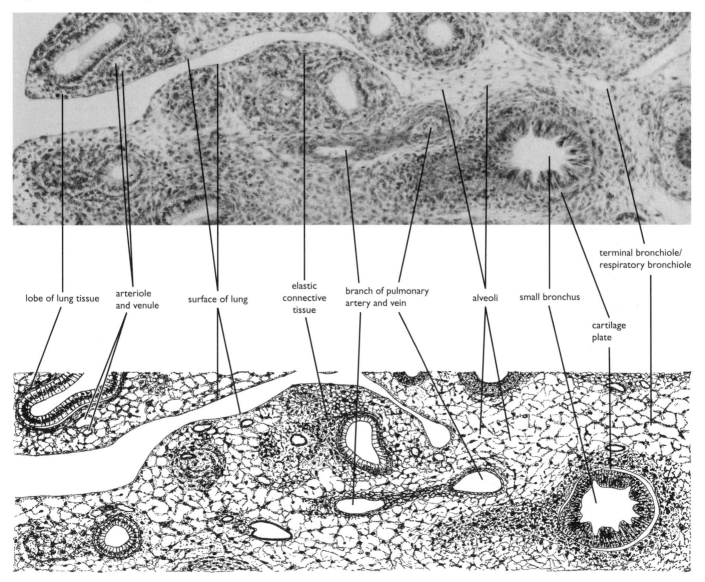

lobe of lung tissue

arteriole and venule

surface of lung

elastic connective tissue

branch of pulmonary artery and vein

alveoli

small bronchus

terminal bronchiole/ respiratory bronchiole

cartilage plate

Capillary network to the alveoli

Networks of capillaries surround the alveoli, supplied by branches of the pulmonary artery and drained by branches of the pulmonary vein. The alveolus wall is lined by squamous epithelium bonded to the capillary wall, and is extremely thin. About 100 cm³ of blood is contained within the capillary networks of the lungs at any one moment. This tiny quantity of blood is spread out, ensuring very rapid exchange of respiratory gases by diffusion. The alveoli are an efficient gaseous exchange system, even though they are a cul-de-sac system with a residual volume of air (page 25) that is diluted with fresh air on each inspiration.

3 What factors affect the rate of diffusion of gases through a surface such as that of the lungs?

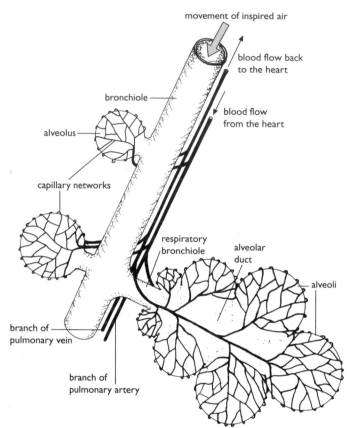

Figure 4.8 The blood supply to the alveoli.

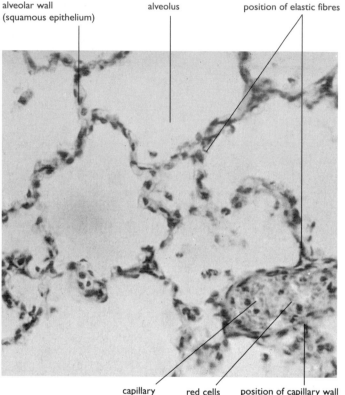

Figure 4.9 TS of alveoli, HP.

The alveoli contain two types of **protective cell**:

1 **dust cells** (macrophages), the main detritus-collecting cells of the body: they originate from the bone marrow stem cells, migrate to the lymph nodes and are then dispersed around the body in the blood circulation (Figure 5.15, page 40). They are amoeboid cells, and they lie in wait, ready to mop up foreign matter. The lung surface is one such site;

2 **surfactant cells**: these produce a phospholipid-rich secretion known as surfactant that lines the inner surface of the alveoli. It lowers the surface tension, permitting the alveoli to flex as the pressure of the thorax falls and rises.

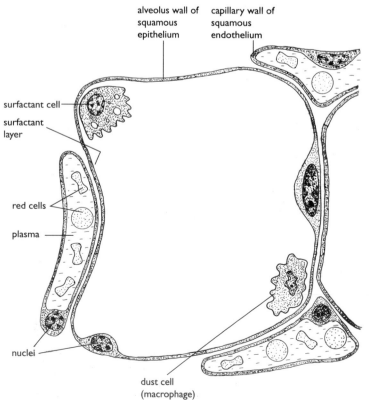

Figure 4.10 The role of macrophages (dust cells) and surfactant cells.

Smoking and the lungs

For many years it has been appreciated that smoking is bad for our health, but only recently has it become clear how bad. The effects of prolonged smoking catch up with us from the age of 35 years onwards. Smokers are much more likely to suffer (and die prematurely) from a list of diseases, including asthma, heart disease, strokes, emphysema, chronic bronchitis, cancers of the mouth, throat, lung, oesophagus, stomach, pancreas, bladder and rectum, and of a type of leukaemia called myeloid leukaemia.

Emphysema is a disease of the respiratory bronchioles and alveoli in which the alveolar walls lose their elasticity and alveoli remain filled with air (fail to expire air properly). Such lungs contain far too much of the natural enzyme that breaks down the elastin of the alveolar walls, and far too little of the natural inhibitor of this enzyme. Therefore, the machinery to repair damaged alveoli is inactivated. Alveoli merge to form huge air spaces, which cause a lowered surface area for gas exchange. The patient becomes permanently breathless.

The disease can be prevented by stopping smoking, but any damage already done to the lungs cannot be reversed.

Healthy lung tissue (x 200)

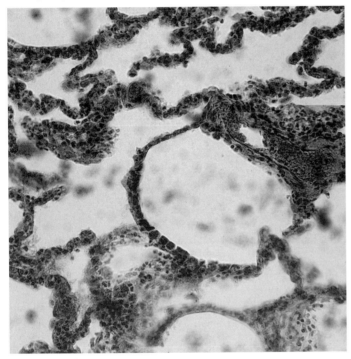

Lung tissue showing advanced emphysema (x 200)

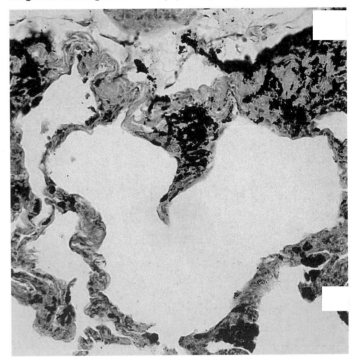

Figure 4.11 Photomicrographs of human lung tissue.

Chronic bronchitis is a disease in which the bronchi become inflamed. The mucous glands and goblet cells of the epithelium enlarge, and the ciliary current becomes ineffective. Excess mucus is secreted. Once contracted, it is a disease that may typically last for 3 months.

Lung cancer. Continuing, persistent exposure of the bronchi to cigarette smoke (and certain other pollutants) causes enlargement of the goblet cells. Basal cells that normally replace goblet cells as they wear out start to divide abnormally. These cells are quite likely to be the seat of a lung cancer called bronchiogenic carcinoma. Heavy cigarette smokers are 20 times more likely to develop this disease than non-smokers.

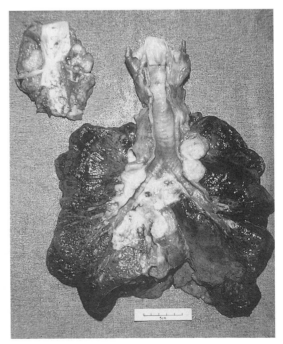

Figure 4.12 Human lung tissue with cancer. These tumours are formed from unspecialised masses of cells that continue to grow, uncontrolled by signals from other cells around them. Lung cancer is the third most common cause of death in the UK.

Blood

Blood is classified as a connective tissue (page 10), even though it has a fluid matrix (the blood plasma), which is not secreted by the blood cells themselves.

Figure 4.13 The composition of the blood.

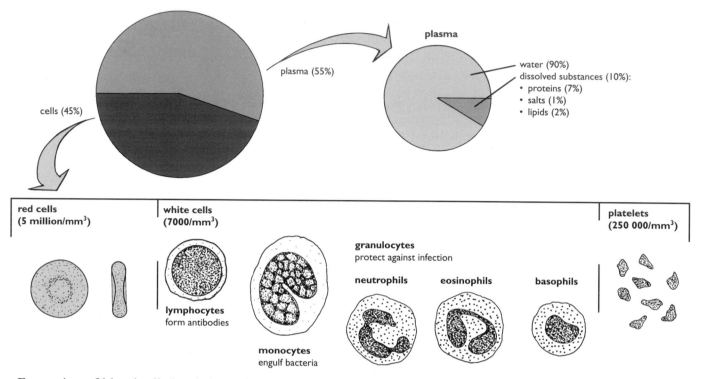

Formation of blood cells (or their precursors) occurs in red bone marrow tissue.

Red cells are formed at a great rate of approximately 200 000 million every day, but their life-span is only about 120 days. As these cells mature they lose their nucleus and their cytoplasm acquires a huge quantity of the conjugated protein, haemoglobin. Breakdown of red cells occurs in the spleen and liver (page 52).

Platelets are tiny disc-shaped cells that also lose their nucleus during formation. They play a key part in blood clotting (page 35).

White cells are of different types, and during development some leave the bone marrow to undergo a maturation phase in the lymph nodes, the skin or the thymus gland. White cells are all concerned with the body's defence against infection (page 40).

Plasma components are diverse, and are contributed by a range of organs. However, the overall composition of plasma is effectively regulated and controlled by the actions of the liver and kidneys.

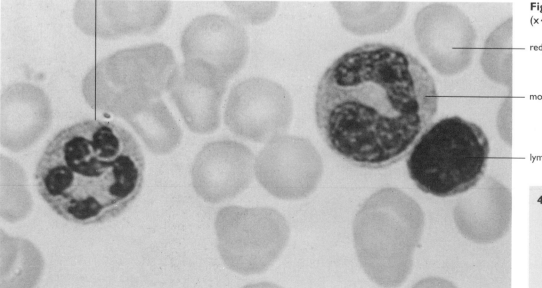

Figure 4.14 Human blood smear (× 4500).

4 One of the major roles of the blood circulation is as the transport system of the body. What does blood transport?

Transport of respiratory gases

Oxygen is only slightly soluble in water (and therefore in the plasma), but each red cell contains about 280 million molecules of haemoglobin. Haemoglobin is a compact molecule, built from four interlocking sub-units. Each sub-unit has an active site to which a molecule of oxygen will bind, provided the partial pressure of oxygen (Po_2) is relatively high, as is the case in the alveolar capillaries where 'loading' occurs. In respiring tissues, the Po_2 is much lower, and under this condition the oxyhaemoglobin 'unloads' its oxygen molecules. The difference in partial pressures of carbon dioxide (Pco_2) in respiring cells (high) and in the lungs (lower) aids the loading/unloading sequence (the Bohr effect).

Figure 4.15 Haemoglobin, its affinity for molecular oxygen molecules, and the effects of Pco_2.

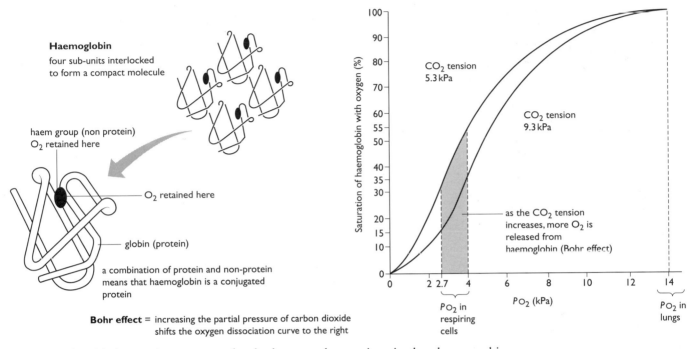

Carbon dioxide is mostly transported as hydrogencarbonate ions in the plasma and in the red cells. Red cells contain the enzyme carbonic anhydrase, which greatly accelerates the formation of hydrogencarbonate ions. The H^+ ions are buffered by the haemoglobin, preventing the blood from becoming acidic (Figure 4.16)

Figure 4.16 A summary of the transport of respiratory gases in the blood.

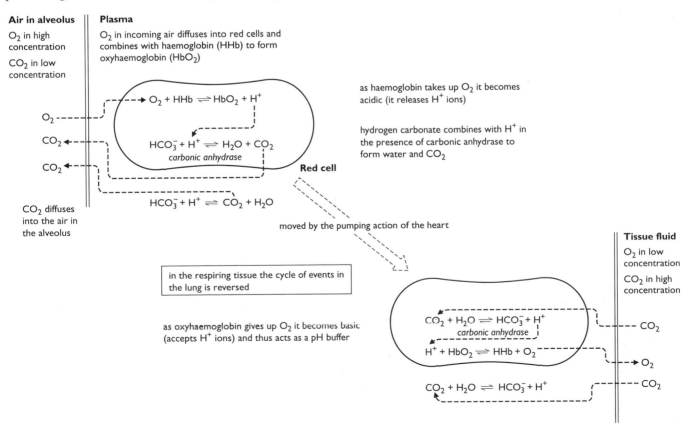

Heart and circulation

5

Mammals have a **'closed' circulation**, that is, blood is contained in vessels (arteries, veins and capillaries) and pumped at pressure by a muscular heart. The mammalian system (and that of birds) is also described as a **'double circulation system'**, in that blood passes twice through the heart in every complete circulation of the body, going to the lungs via the pulmonary system, and then to the rest of the body via the systemic system. By contrast, fish have a single circulation system (from the heart to the gills to the body, and back again). The advantages of a double circulation are:

- it allows simultaneous high-pressure delivery of oxygenated blood to all regions of the body;
- it ensures that oxygenated blood reaches the respiring tissues, undiluted by deoxygenated blood.

1 Insects are said to have an 'open' circulation. What does this mean?

2 What distinguishes a portal vein?

In the **mammalian circulation** each organ is supplied with blood from the heart by an artery that branches from the main aorta. Arteries branch profusely within the organ, forming numerous arterioles supplying the capillaries among the cells. All cells are very close to capillaries, and cells are bathed in tissue fluid formed from the blood. The branching/return sequence is:

artery
↓
arteriole
↓
capillary
↓
venule
↓
vein

and the veins eventually join a vena cava, carrying blood back to the heart.

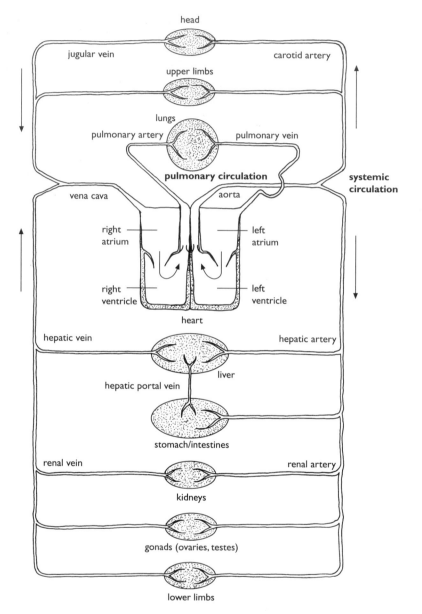

Figure 5.1 The layout of the mammalian blood circulation.

Artery and vein

The walls of the arteries and veins have three layers, but the walls of arteries are much thicker than those of veins. The innermost layer is of squamous epithelium (page 10). This also lines the heart and lymph vessels, and forms the walls of the capillaries (page 34). The middle layer is of smooth muscle and elastic fibres (tunica media). Veins have a much thinner tunica media than arteries. The outer layer is of fibrous connective tissue (tunica externa). Blood in veins is under low pressure, and backflow is prevented by semi-lunar valves.

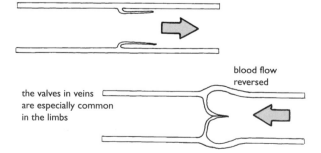

the valves in veins are especially common in the limbs

blood flow reversed

These valves are similar in structure to the semi-lunar valves of the heart

Figure 5.2 Valves in veins.

Figure 5.3 The structure of an artery and a vein.

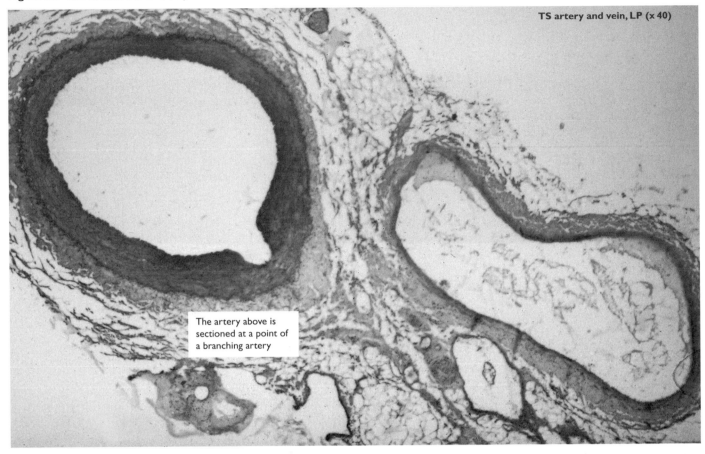

TS artery and vein, LP (x 40)

The artery above is sectioned at a point of a branching artery

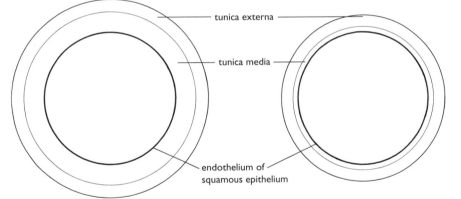

tunica externa

tunica media

endothelium of squamous epithelium

The thin walls of veins mean that blood in them is more visible through the skin than in thick-walled arteries

Table 5.1 Differences between arteries, veins and capillaries.

	Artery	Vein	Capillary
Structural differences Tunica externa (collagen fibres)	thick layer	thin layer	absent
Tunica media (elastic fibres and involuntary muscle fibres)	thick layer	thin layer	absent
Endothelium (squamous epithelium)	present	present	present
Valves	absent	present	absent
Functional differences	transports blood away from the heart; under high pressure; blood flows rapidly; blood moves in pulses	transports blood towards the heart; under low pressure	site of exchange between blood and tissues; falling pressure

Structure and function of capillaries

All body tissues are supplied by capillary networks (exceptions are cartilage and the cornea of the eye). Muscles are particularly richly supplied. The branching of capillaries and their thin walls facilitate exchange between blood and tissue fluid. In some specialised situations, including the glomeruli of the kidney tubules (page 44), there are also systems of tiny pores through the endothelial cytoplasm.

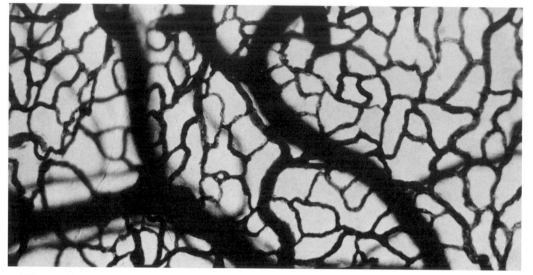

Figure 5.4 Capillary network in vertebrate skin (x 30).

The skin capillary network can be observed in the taut skin of the webbed feet of the frog (*Rana*).

Tissue fluid is the means by which nutrients are delivered to the cells, and waste products and secretions are removed. Many components of plasma escape through the wall of the capillaries to form tissue fluid, but none of the red cells and almost none of the soluble plasma proteins escape.

Figure 5.5 Exchange between blood and cells via tissue fluid.

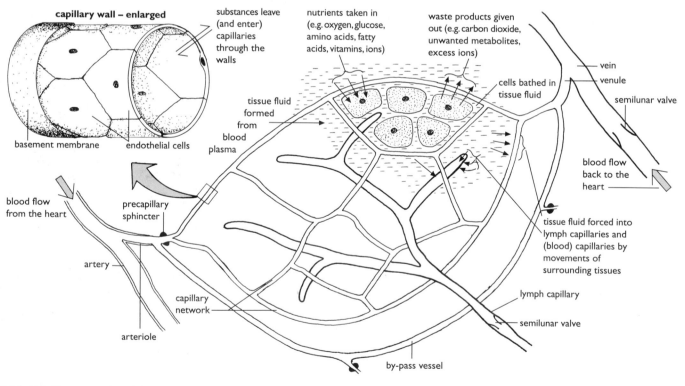

Table 5.2 Transport roles of the blood circulation.

In **tissue respiration**: oxygen to all the tissues, and carbon dioxide back to the lungs

In **hydration**: transporting water to all the tissues

In **nutrition**: nutrients, organic (sugars, amino acids, lipids and vitamins) and inorganic ions (e.g. K^+, Ca^{2+}), to all cells

In **excretion**: transporting waste products of metabolism to kidneys, lungs, sweat glands, liver

In **temperature regulation**: by distribution of heat

In **development and coordination**: transporting hormones from endocrine glands to target organs all over the body

Formation of tissue fluid occurs due to the forces of blood pressure (hydrostatic pressure) and diffusion (remember that osmosis is a special case of diffusion). The net changes are that water and nutrients flow out of the capillaries at the arterial end of the capillary bed, and water with waste products (and some nutrients) flows back into the capillary at the venous end. More fluid flows out than in, and the remainder enters the lymph system. Lymph vessels have tiny entry pores (one-way valves), and lymph also carries away anything too large to pass back through the capillary walls.

> **3** How will the composition of tissue fluid typically differ from that of the plasma?

Figure 5.6 The working capillary network.

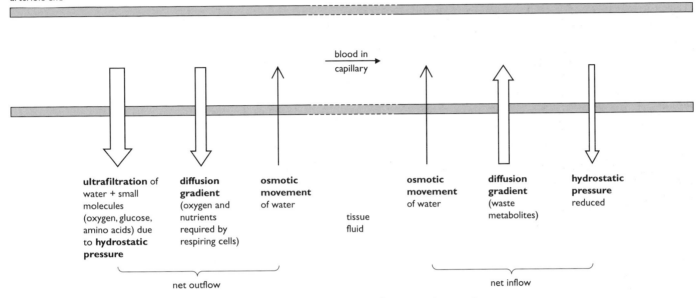

The clotting mechanism copes with leaks and breakages, either at an internal haemorrhage or at a skin surface cut. The normal reaction is for the break to be sealed quickly.

Figure 5.7 Blood clot formation.

To stop bleeding: i) constriction of the vessel, ii) aggregation of platelets, and iii) formation of a fibrin clot occurs. These steps are regulated by enzymes functioning in a cascade model, each released from an inactive (precursor) form. The mechanism ensures that no accidental triggering occurs; clotting is initiated by conditions at the damage site

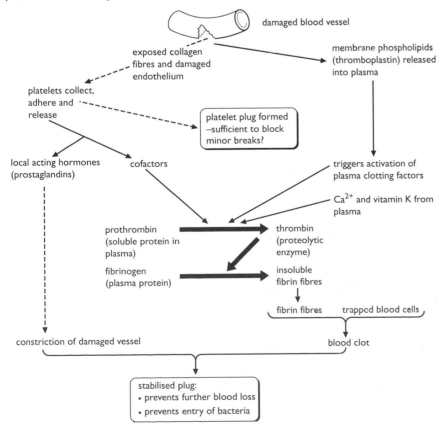

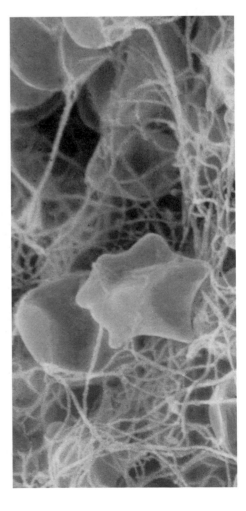

The heart

The human heart, a muscular organ about the size of a clenched fist, is divided into four chambers. Direction of blood flow through the heart is maintained by valves. The chambers of the right side are completely separated from those of the left side. The two upper chambers (atria) are comparatively thin walled; the two lower chambers (ventricles) are thick walled. The muscular wall of the left ventricle is about three times as thick as that of the right ventricle, but the volumes of the left and right sides (the quantities of blood they contain) are identical.

in situ, the heart is surrounded and contained in a tough, non-elastic membrane, the **pericardium**

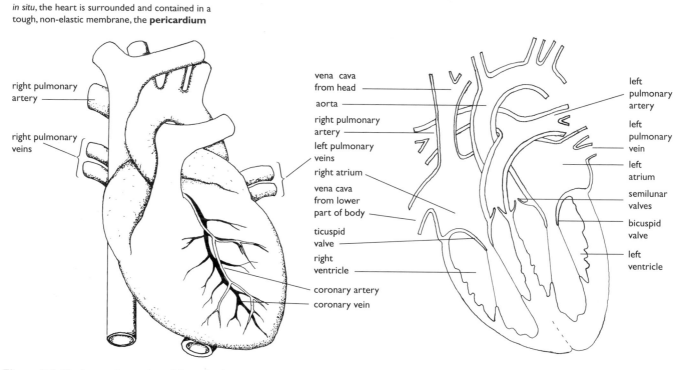

Figure 5.8 The human heart viewed from the front of the body showing external appearance, and in LS.

The **muscular wall of the heart** is supplied with oxygenated blood by coronary arteries from the aorta. Heart muscle (cardiac muscle) consists of cylindrical, branching columns of muscle fibres with special electrical junctions called intercalated discs. In electronmicrographs, cardiac muscle fibres are seen to be surrounded by sarcolemma (muscle plasma membrane), and many mitochondria are present.

4 Cardiac muscle fibres are also described as 'coencytic'. What does this mean?

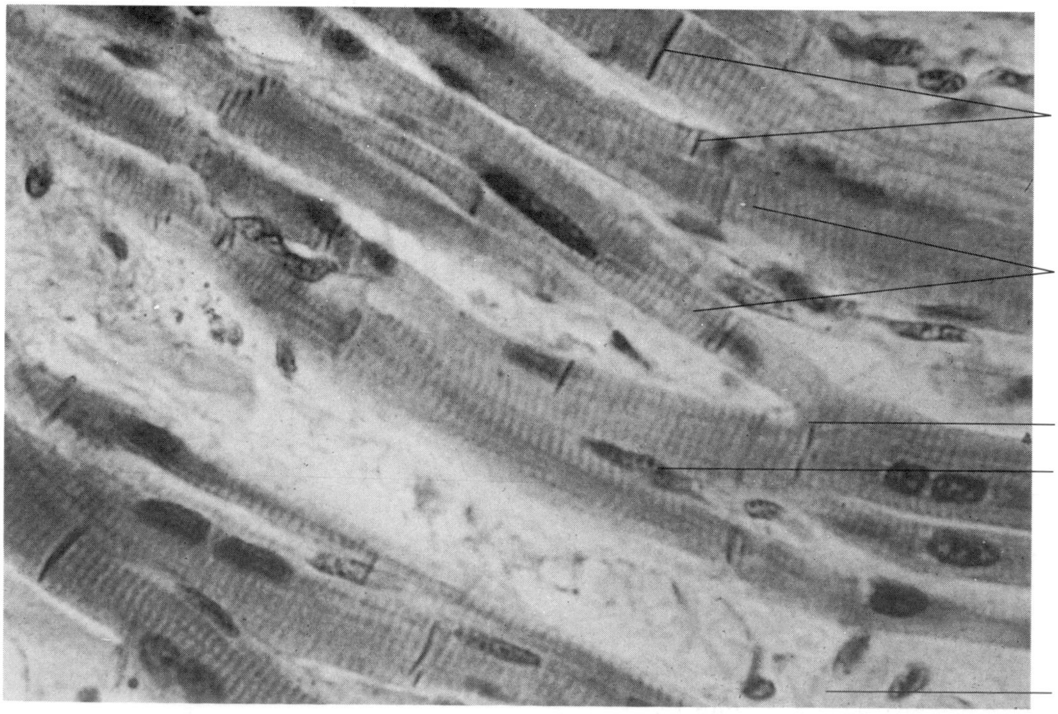

Figure 5.9 Cardiac muscle, HP.

intercalated discs

striations

branch

nucleus

capillaries lie between the fibres

The **pumping action of the heart** consists of alternate contractions (**systole**) and relaxations (**diastole**). Contraction of cardiac muscle is followed by relaxation and elastic recoil of the heart, because of the elastic connective tissue present alongside the cardiac muscle fibres. The resulting changing pressure of blood in atria, ventricles, pulmonary artery and aorta automatically opens and closes the valves. This cycle of events, the **cardiac cycle**, typically lasts about 0.8 seconds (75 beats/minute), but the heart rate responds to a number of factors (page 38).

Figure 5.10 The cardiac cycle and the associated pressure changes.

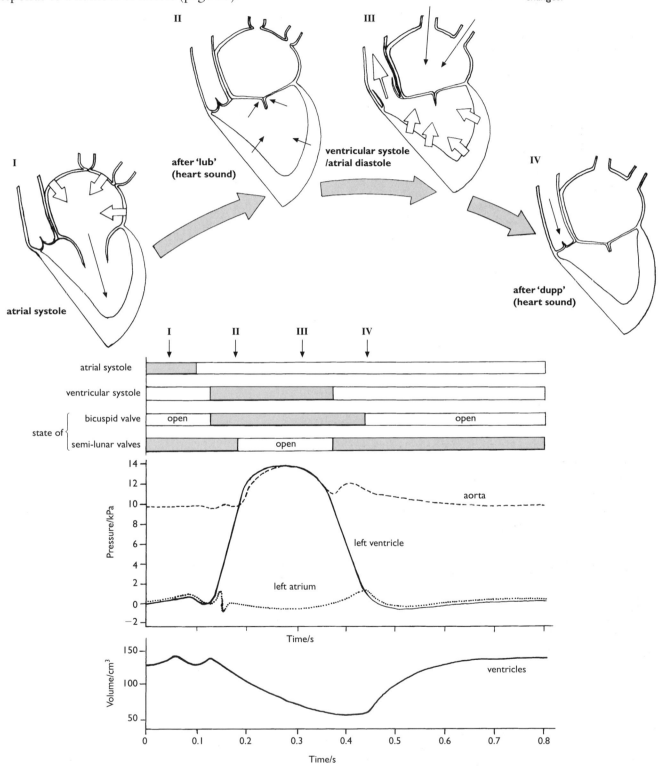

Heart sounds are detected by the use of a stethoscope. The valves of the heart close passively as soon as there is a tendency for blood to flow in the reverse direction. In a healthy heart, as the atrio-ventricular (tricuspid and bicuspid) valves close a 'lub' is heard, and as the semi-lunar valves close a 'dupp' is heard.

Heart 'sounds' provide information about the valves. Unusual sounds are called murmurs, and (for example) may indicate incomplete closure of the atrio-ventricular (tricuspid and/or bicuspid) valves, allowing backflow of some blood.

Control of the heartbeat

The origin of the heartbeat is from within the heart muscle itself (**myogenic** origin), starting in the sino-atrial node or **pacemaker** in the wall of the right atrium. Muscle fibres conduct excitations to both atria, causing contraction of the muscular walls. The ventricles are relaxed at this stage. Next, the atrio-ventricular node at the base of the right atrium picks up the excitation and passes it to the ventricles via modified muscle fibres called Purkinje fibres. The ventricles then contract, whilst the atria relax. After contraction, all muscle fibres are insensitive to fresh stimulation for a very brief period (the refractory period). In cardiac muscle the refractory period is prolonged, helping it to contract forcefully, without seizing up (tetanus).

Figure 5.11 Myogenic stimulation of cardiac muscle.

Cardiac muscle, HP

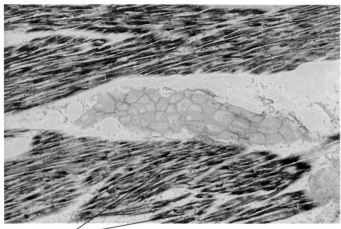

Purkinje fibres

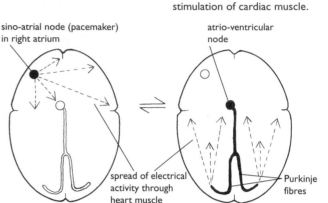

sino-atrial node (pacemaker) in right atrium

atrio-ventricular node

spread of electrical activity through heart muscle

Purkinje fibres

Impulse transmission through the above conduction system generates electrical currents that can be detected on the body surface, and may be recorded as an **electrocardiogram**

Cardiac output = volume of blood pumped per beat × number of beats per minute.

In practice, cardiac output is regulated by three mechanisms:

1   Cardiac muscle responds to **stretching** by contracting more strongly. An increased volume of blood is pumped out e.g. during heavy physical exercise, the flow of blood to the gut is typically reduced, causing enhanced flow to organs such as the skeletal muscles. As a result, blood flow back to the heart is enhanced and the heart is slightly distended.

2   The action of the pacemaker is regulated by **involuntary reflex action**. Stretch receptors located in the walls of the aorta, carotid arteries and vena cavae detect changes in blood pressure, and impulses are sent via the cardiovascular centre in the medulla of the hindbrain to the pacemaker. When blood pressure is high in the arteries, the rate of heartbeat is slowed; when it is low, the rate of heartbeat (and force of ventricular contraction) is increased.

3   At times of stress, the **hormone** adrenaline is secreted by the adrenal gland. At the pacemaker adrenaline causes the heartbeat rate to increase.

Figure 5.12 How cardiac output is regulated.

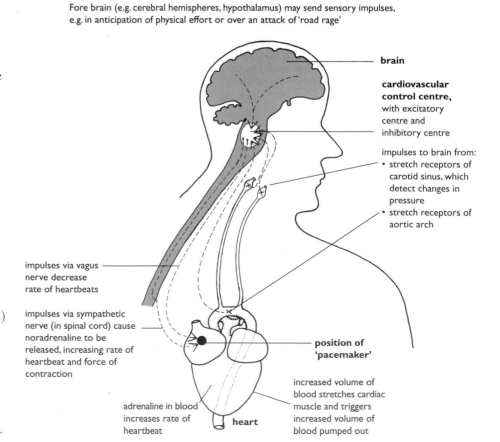

Fore brain (e.g. cerebral hemispheres, hypothalamus) may send sensory impulses, e.g. in anticipation of physical effort or over an attack of 'road rage'

brain

cardiovascular control centre, with excitatory centre and inhibitory centre

impulses to brain from:
• stretch receptors of carotid sinus, which detect changes in pressure
• stretch receptors of aortic arch

impulses via vagus nerve decrease rate of heartbeats

impulses via sympathetic nerve (in spinal cord) cause noradrenaline to be released, increasing rate of heartbeat and force of contraction

position of 'pacemaker'

increased volume of blood stretches cardiac muscle and triggers increased volume of blood pumped out

adrenaline in blood increases rate of heartbeat

heart

Aspects of cardiovascular disease

1 The natural **pacemaker** may become defective. Battery-powered pacemakers may be implanted under the skin of the thorax wall, with electrodes to the atrial walls introduced intravenously. In some designs the pacemaker is sensitive to any natural heartbeats; an impulse is delivered only when the heart misses a beat.

2 Degeneration of an artery wall may arise from deposition of fatty material (an **atheroma**) under the endothelium. The deposit builds up from cholesterol in the blood, carried around the body as 'low-density lipoproteins'. Dense white fibrous tissue may also form and the endothelium breaks down, allowing the blood flow to come into contact with fatty and fibrous tissues. Platelets tend to stick at such sites, and conditions then favour the formation of a blood clot – a **thrombus**.

fatty material deposited

restricted blood flow

Figure 5.13 Artery damaged by atherosclerosis.

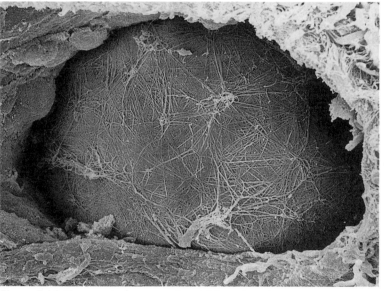

Figure 5.14 Human pulmonary vessel containing a thrombus.

3 Blood pressure is measured in the arm by a sphygmomanometer. Normally, systolic and diastolic blood pressures are about 15.8 and 10.5 kPa (120 and 70–80 mmHg), respectively. However, blood pressure tends to rise with age, with obesity, and with experience of persistent stress and anxiety. **Raised blood pressure** is a risk factor in heart disease. Drugs called beta-blockers are prescribed because they inhibit the sympathetic nerves (by antagonising the action of noradrenaline), and lower blood pressure.

4 **Atherosclerosis** may lead to the weakening of an artery wall to the point where blood pressure seriously stretches the wall. This localised bulge is called an **aneurysm**, and may burst at any time.

5 A blood clot (**thrombus**) formed at an atherosclerotic site may break away and be carried around the circulation (as an **embolus**). If the embolus lodges in a narrower artery it obstructs the blood supply locally (an **embolism**).

Blood clots formed at an atherosclerotic site may cause a blockage there (a **thrombosis**). If the obstruction occurs in a coronary artery, a region of heart muscle is deprived of oxygen. Initially the patient experiences chest pain, known as **angina**. If the blockage is maintained and the condition not treated, the area of heart muscle dies from lack of oxygen. A heart attack (or **myocardial infarction**) has occurred. With death of heart tissue there is a danger that an irregular heartbeat rhythm may arise (**ventricular fibrillation**), and the heart may ultimately cease to be an effective pump.

Coronary arteries damaged by atherosclerosis can be surgically by-passed (by a **coronary by-pass operation**) with short lengths of artery or vein taken from other parts of the patient's body. If a thrombosis or embolism blocks an artery supplying part of the brain, a **'stroke'** is the outcome.

5 Why is brain tissue especially vulnerable to interruption of the blood supply?

The battle against infection

The blood circulation plays a part in the resistance to infection when the barriers of external skin or internal epithelia are breached. A response to cuts and abrasions is **inflammation**, a process by which blood, tissue fluid and then white cells accumulate at the site of damage and assist in the removal of microorganisms and the repair of the tissues. If a blood vessel is ruptured, then the **clotting** mechanism is activated (page 35).

White cells originate in the bone marrow by division of the stem cells from which the red cells form. However, in development, white cells migrate elsewhere in the body and they retain their nucleus. White cells fall into two functional groupings.

6 What are lysosomes and how do they effectively destroy ingested bacteria?

1 White cells that engulf 'foreign' material (phagocytosis)

Neutrophils are short-lived. They ingest bacteria within the blood circulation or migrate outside the capillaries to act at a wound site.

Monocytes are long-living phagocytic cells that leave the blood circulation and lie in wait in organs, such as lungs (page 28), liver (page 48), spleen and kidneys. They are the principal 'rubbish-collecting' cells or macrophages of the body.

Figure 5.15 Lymph nodes and the formation of phagocytic white cells.

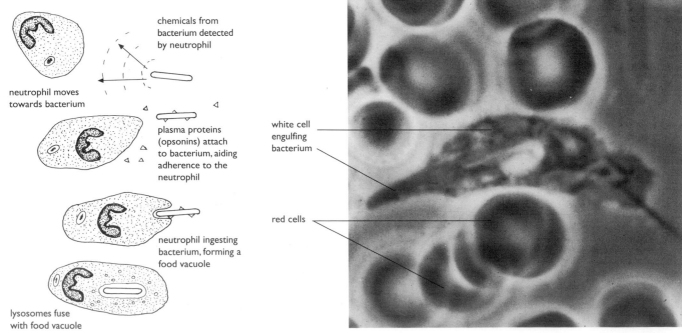

Formation of phagocytic white cells

stem cell of the bone marrow

cells divide to form phagocytic white cells (and red cells)

white cells migrate to tissues

short-lived cells   long-lived cells

monocytes retained in lymph nodes

divide to form macrophages

Venue of phagocytic white cells
Lymph capillaries form a network in the body tissues. Lymph nodes are part of this network, and lymph passes through nodes before draining back into the blood circulation

here white cells engulf debris and pathogens

lymph node

lymphatic

The lymphatic system is part of the circulation system, but is also part of the body's defences

(white cells forming antibodies also occur here, see Figure 5.17)

lymph capillary

debris passes via valve-like pores

vein

body tissue

tissue fluid passes back into capillaries

heart   artery

blood capillary

tissue fluid with metabolites

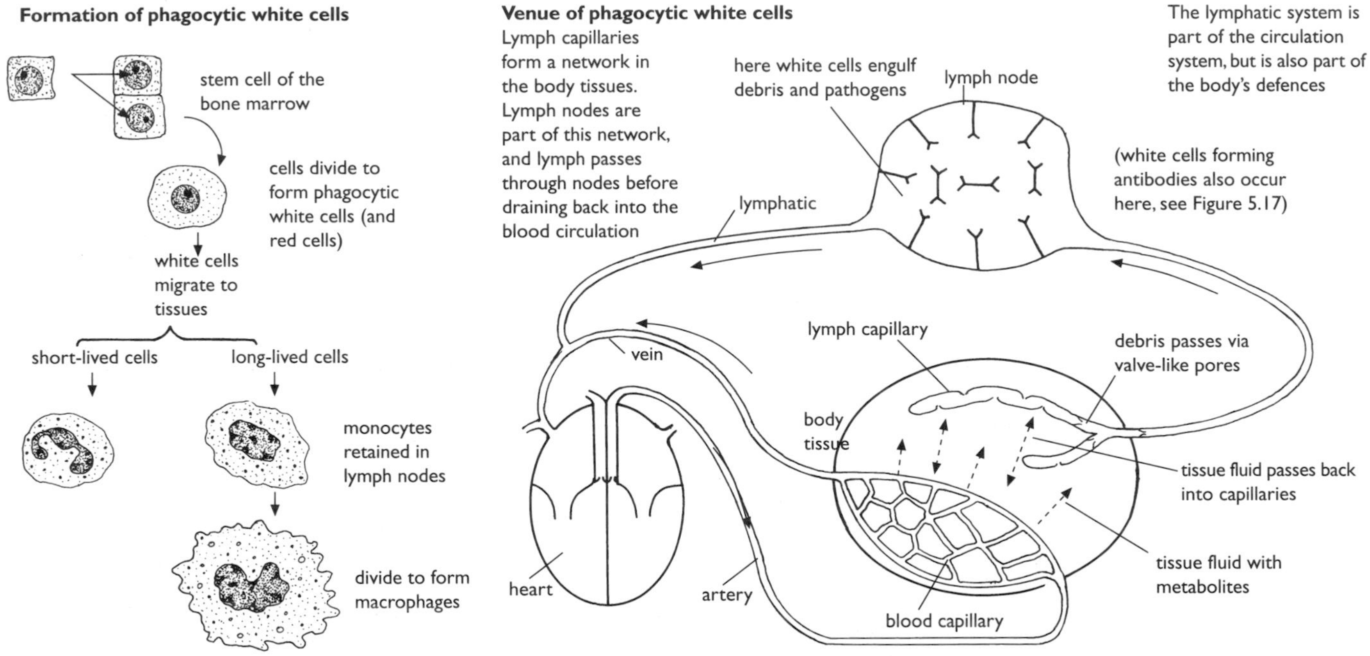

Figure 5.16 Neutrophils engulfing bacteria in the blood.

chemicals from bacterium detected by neutrophil

neutrophil moves towards bacterium

plasma proteins (opsonins) attach to bacterium, aiding adherence to the neutrophil

neutrophil ingesting bacterium, forming a food vacuole

lysosomes fuse with food vacuole

Magnification (x 6000)

white cell engulfing bacterium

red cells

2 Immune system white cells, involved in specific resistance to disease

The immune system is based on the body's ability to recognise 'self' (e.g. body cells and specific proteins) and to tell them apart from **antigens**, that is, 'non-self' substances typically produced by an invading organism. This ability to recognise antigens and to take steps to overcome them is the property of a type of white cell called **lymphocytes**. All lymphocytes originate from stem cells in the bone marrow.

B-lymphocytes (bone-derived lymphocytes) migrate directly from bone marrow to lymph nodes during development. **T-lymphocytes** (thymus-derived lymphocytes) migrate from the bone marrow of the fetus to the thymus gland to mature. Later, the mature T-lymphocytes migrate to the lymph nodes. Both B- and T- lymphocytes are activated by the presence of an antigen and they respond in specialised ways to protect the body systems.

Figure 5.17 Formation and roles of the immune system white cells (B- and T- lymphocytes) – a summary.

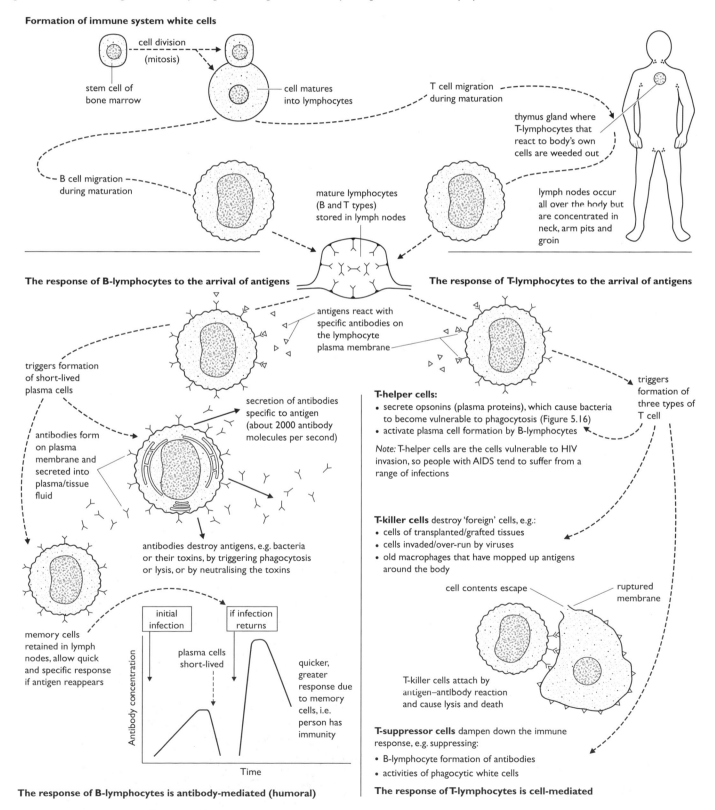

Formation of immune system white cells

cell division (mitosis)

stem cell of bone marrow

cell matures into lymphocytes

T cell migration during maturation

thymus gland where T-lymphocytes that react to body's own cells are weeded out

B cell migration during maturation

mature lymphocytes (B and T types) stored in lymph nodes

lymph nodes occur all over the body but are concentrated in neck, arm pits and groin

The response of B-lymphocytes to the arrival of antigens

antigens react with specific antibodies on the lymphocyte plasma membrane

The response of T-lymphocytes to the arrival of antigens

triggers formation of short-lived plasma cells

antibodies form on plasma membrane and secreted into plasma/tissue fluid

secretion of antibodies specific to antigen (about 2000 antibody molecules per second)

antibodies destroy antigens, e.g. bacteria or their toxins, by triggering phagocytosis or lysis, or by neutralising the toxins

triggers formation of three types of T cell

T-helper cells:
- secrete opsonins (plasma proteins), which cause bacteria to become vulnerable to phagocytosis (Figure 5.16)
- activate plasma cell formation by B-lymphocytes

Note: T-helper cells are the cells vulnerable to HIV invasion, so people with AIDS tend to suffer from a range of infections

T-killer cells destroy 'foreign' cells, e.g.:
- cells of transplanted/grafted tissues
- cells invaded/over-run by viruses
- old macrophages that have mopped up antigens around the body

cell contents escape

ruptured membrane

memory cells retained in lymph nodes, allow quick and specific response if antigen reappears

initial infection

if infection returns

plasma cells short-lived

quicker, greater response due to memory cells, i.e. person has immunity

Antibody concentration

Time

T-killer cells attach by antigen–antibody reaction and cause lysis and death

T-suppressor cells dampen down the immune response, e.g. suppressing:
- B-lymphocyte formation of antibodies
- activities of phagocytic white cells

The response of B-lymphocytes is antibody-mediated (humoral)

The response of T-lymphocytes is cell-mediated

Excretion and osmoregulation

6

Excretion is the removal from the organism of the waste products of metabolism together with excess salts ingested and toxic substances. The important excretory products of mammals are nitrogenous waste (urea) and carbon dioxide. Urea is excreted at the kidneys, carbon dioxide at the lungs (page 25).

Osmoregulation is the process by which the balance of water and dissolved solutes is regulated. Osmoregulation and excretion are closely connected. In mammals, the kidneys are where most of the excess water and dissolved substances are removed from the blood (some water and solutes are also lost at the skin in mammals that sweat, see page 54). Dissolved substances include non-electrolytes such as urea, and electrolytes such as inorganic ions (e.g. K^+, Na^+).

The human **kidneys** are about the size of a clenched fist, and lie attached to the dorsal wall of the abdominal cavity, behind the liver. In longitudinal section the kidneys show an outer **cortex** and an inner **medulla**. A huge number of tiny tubules called **nephrons** are present in each kidney, typically with part of each nephron in the cortex and part in the medulla. The nephron is well supplied by capillaries and is the site of osmoregulation and excretion.

1 Nitrogenous excretory products are efficiently removed from the body. Why is it just as important that excess salts and water are removed speedily?

Figure 6.1 The mammalian kidneys.

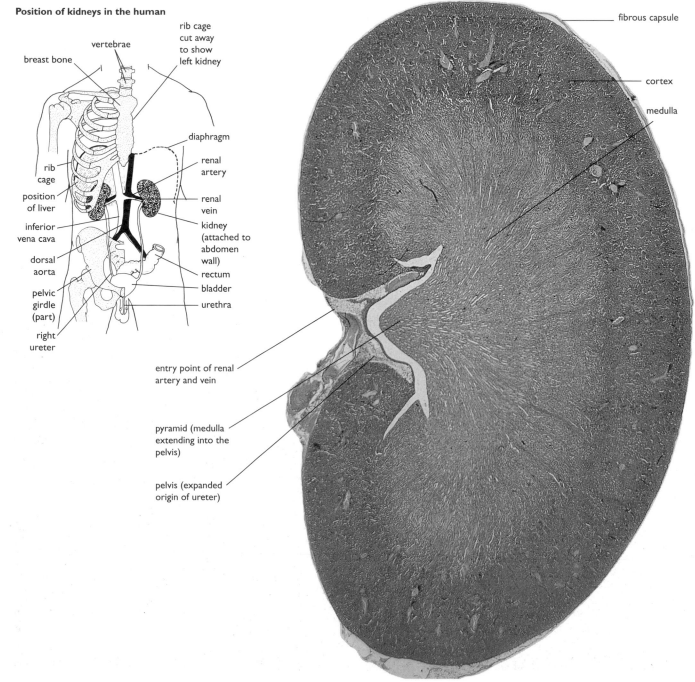

Position of kidneys in the human

- breast bone
- vertebrae
- rib cage cut away to show left kidney
- rib cage
- position of liver
- inferior vena cava
- dorsal aorta
- pelvic girdle (part)
- right ureter
- diaphragm
- renal artery
- renal vein
- kidney (attached to abdomen wall)
- rectum
- bladder
- urethra

LS of mammalian kidney, LP

- fibrous capsule
- cortex
- medulla
- entry point of renal artery and vein
- pyramid (medulla extending into the pelvis)
- pelvis (expanded origin of ureter)

A **nephron** is a tiny tubule with thin walls, about 3 cm in length, composed of five functional regions:

- Malpighian body, consisting of blood capillaries (the glomerulus) surrounded by the renal capsule;
- proximal convoluted tubule;
- descending and ascending limbs of the loop of Henle;
- distal convoluted tubule;
- collecting duct.

The blood supply is an integral part: beginning at the arteriole serving the glomerulus, the blood is then carried either to the capillary network around the convoluted tubules, or as a single capillary (the vasa recta) running alongside the loop of Henle.

There are **two types of nephrons**, largely based on their positions in the kidneys. **Cortical** nephrons occur largely in the cortex, with a short loop of Henle extending just into the medulla. **Juxtamedullary** nephrons have long loops of Henle extending deep into the medulla. Aquatic mammals contain cortical nephrons only, desert mammals have juxtamedullary nephrons only, but most mammals have a mixture of various proportions.

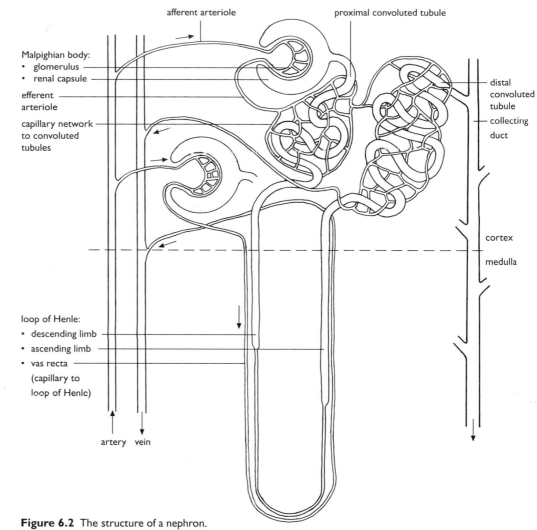

Figure 6.2 The structure of a nephron.

Figure 6.3 Cortex tissue of the kidneys, TS.

Showing renal capsules

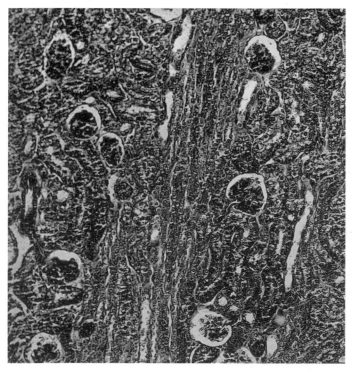

Injected to show blood supply to glomeruli

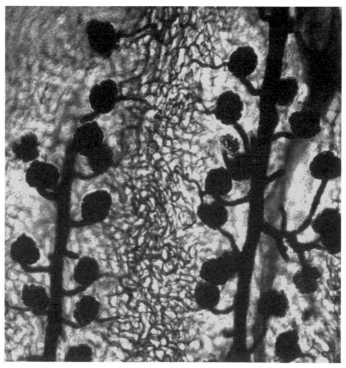

The working kidney – steps to urine formation

Approximately 1.0–1.5 litres of urine are produced per day. This contains about 50 g solids, of which urea (30 g) and sodium chloride (15 g) make up the bulk. The operation of the nephron involves five distinct stages:

1 **Ultrafiltration in the Malpighian body:** in the glomerulus, water and other small molecules in the plasma are forced out (ultrafiltration) through the walls of the capillaries and renal capsule into the lumen of the nephron. Powered by the pressure of the blood, and made possible by the sieve-like structure of capillary and renal capsule walls, both useful substances and excretory substances are filtered out. Blood cells and most blood protein cannot pass through.

> **2** Urea is produced in the liver from ammonia and carbon dioxide. What is the origin of the ammonia and why is it better handled in the mammal's body as urea?

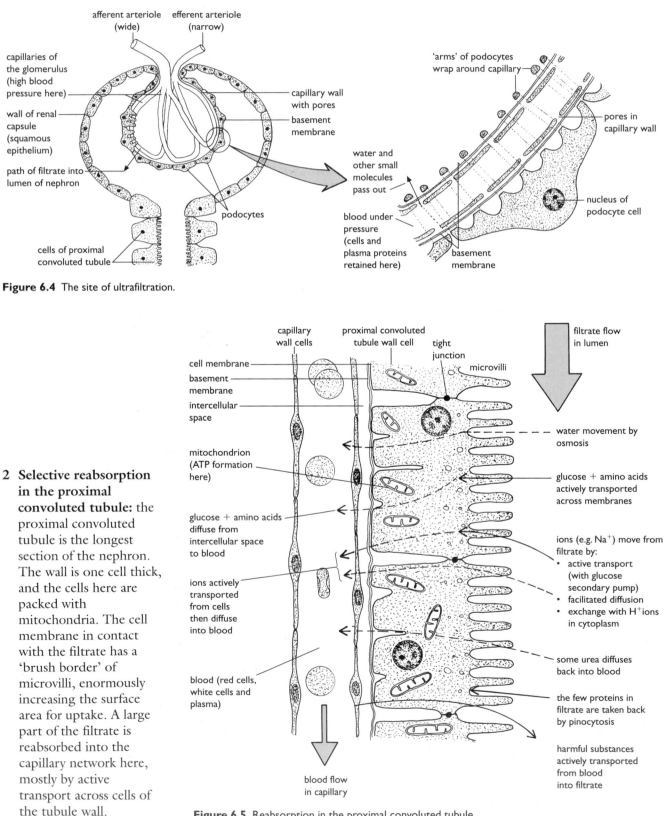

Figure 6.4 The site of ultrafiltration.

2 **Selective reabsorption in the proximal convoluted tubule:** the proximal convoluted tubule is the longest section of the nephron. The wall is one cell thick, and the cells here are packed with mitochondria. The cell membrane in contact with the filtrate has a 'brush border' of microvilli, enormously increasing the surface area for uptake. A large part of the filtrate is reabsorbed into the capillary network here, mostly by active transport across cells of the tubule wall.

Figure 6.5 Reabsorption in the proximal convoluted tubule.

3 Water conservation in the loop of Henle: water loss in excretion is inevitable, but mammals (and birds) are capable of forming urine that is more concentrated than the blood, thereby minimising loss of water. This is possible because the nephron has a loop of Henle, which has two regions – a descending limb and an ascending limb – together with a parallel blood capillary system, the vasa recta. Part of the loop of Henle lies in the medulla, along with the collecting ducts. The loop of Henle is said to function as a counter current multiplier. A high concentration of salts is formed in the medulla, which allows a lot of water to be absorbed from the collecting ducts.

Figure 6.6 Loops of Henle and collecting ducts.

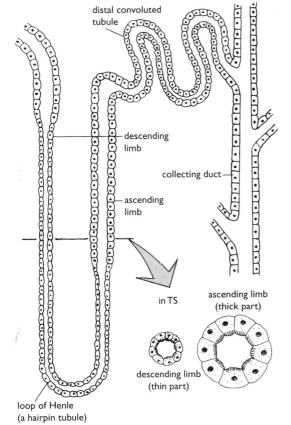

Medulla of kidney in TS (× 800)

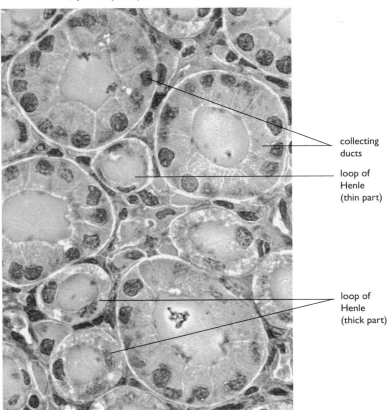

Figure 6.7 The functioning loop of Henle.

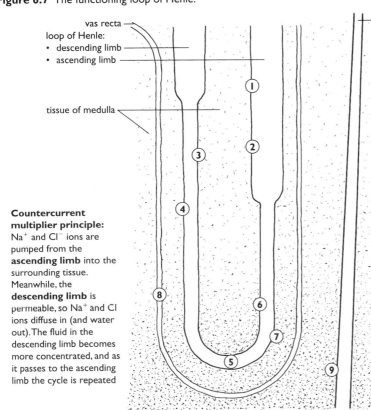

The activities of the tubules are described in the ascending limb first, then in the descending limb, ascending limb and vasa recta

1 here the walls are permeable to ions (Na^+, Cl^-), but impermeable to water
2 here Na^+ and Cl^- are actively transported out
3 the walls of the descending limb are permeable to ions and water
4 here the rising concentration of ions outside causes loss of water by osmosis
5 water loss causes the solution of ions to become concentrated here
6 here Na^+ and Cl^- diffuse out from this concentrated solution
7 more water leaves by osmosis because of the concentration of ions outside – **most** is lost from the (adjacent) collecting ducts
8 the vasa recta deliver oxygen to the cells of the tubule walls
9 there is a gradient in concentration of ions across the medulla, with highest concentration here where all the collecting ducts pass through

Countercurrent multiplier principle: Na^+ and Cl^- ions are pumped from the **ascending limb** into the surrounding tissue. Meanwhile, the **descending limb** is permeable, so Na^+ and Cl ions diffuse in (and water out). The fluid in the descending limb becomes more concentrated, and as it passes to the ascending limb the cycle is repeated

4 **Blood pH and ion concentration regulation in the distal convoluted tubule:** cells of the distal convoluted tubule walls look identical to those of the proximal convoluted tube, but the distal tubules are concerned with:

a) **pH**: the pH of the blood is held within the range pH 7.35–7.45. The pH of the blood tends to **fall** (i.e. H^+ ions accumulate in the blood) when:

- strenuous exercise is prolonged (formation of lactic acid);
- a high protein diet is maintained (sulphur in protein is oxidised to sulphuric acid);
- quantities of acid (e.g. vinegar) are ingested;
- diabetes mellitus is untreated (keto acids accumulate from partial oxidation of fats).

The pH of the blood tends to **rise** (i.e. OH^- ions accumulate in the blood) when:
- quantities of alkali (e.g. $NaHCO_3$) are ingested;
- a vegetarian diet is maintained, because salts of organic acids produce basic products;
- exceptionally, through loss of acid gastric juice by repeated vomiting.

Abrupt changes in blood pH are buffered by plasma proteins, together with hydrogencarbonate and phosphate ions. Longer term adjustments are made in the distal convoluted tubule, as shown in Figure 6.8 below.

3 In what ways are the ions Na^+, Ca^{2+}, K^+ and Mg^{2+} important in the metabolism of mammals, yet NO_3^- is not (unlike in green plants)?

Figure 6.8 Blood pH regulation by the distal convoluted tubule cells.

If blood pH falls below 7.4, H^+ ions from the plasma are secreted in the urine:

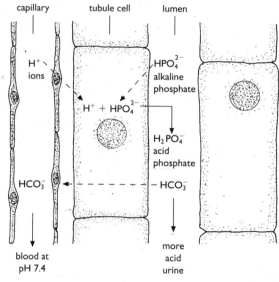

If blood pH rises above 7.4, OH^- ions from the plasma are secreted in the urine:

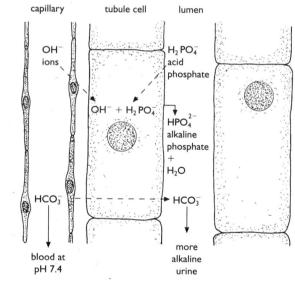

consequently, urine varies between pH 4.5 and 8.2

b) **Plasma ions**: selective reabsorption of ions, typically Ca^{2+}, Mg^{2+}, K^+, Cl^- and especially Na^+, occurs here. Aldosterone from the cortex of the adrenal gland (page 71) maintains a constant level of blood Na^+.

5 **Water reabsorption in the collection ducts – the site of ADH activity**: the permeability of the walls of the collecting duct is variable. The permeability (a case of facilitated diffusion, page 8) is controlled by antidiuretic hormone (ADH) secreted by the pituitary.

a) **When the body is dehydrating**, that is, when very little water is ingested, or we sweat copiously, or excess salt is ingested, the body needs to form a minimum quantity of very concentrated urine. This condition is detected in the hypothalamus (page 69), and ADH secretion is triggered from the posterior pituitary, into the blood.

Figure 6.9 ADH and the production of concentrated urine.

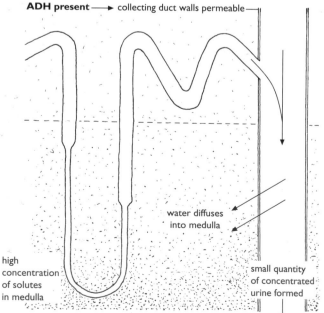

With ADH present, the walls of the collecting ducts are made permeable to water. As a result, and because of the high concentration of salts in the medulla (page 46), reabsorption of water from the lumen of the collecting ducts into the medullary tissue occurs. This water is then redistributed to the body by the capillaries, and only a little concentrated urine is formed.

b) **When excess water is drunk**, that is, when intake exceeds requirements, the solute potential of the blood falls, and the blood volume (and therefore blood pressure) increases. The body's need is for diuresis; copious, dilute urine is formed. This condition is detected in the hypothalamus, and no ADH is secreted into the blood. With ADH absent from the blood, the walls of the collecting ducts become far less permeable, and dilute urine is formed.

Figure 6.11 LS of collecting ducts, HP (× 400).

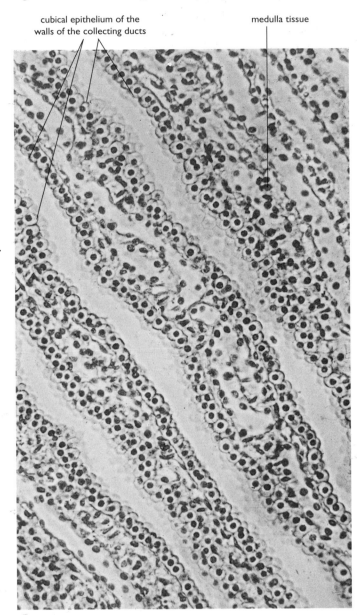

Figure 6.10 The collecting duct and diuresis.

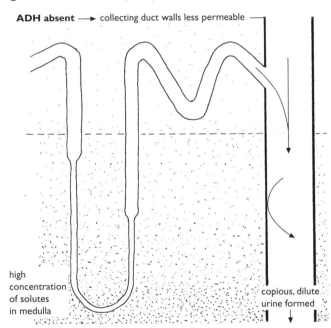

Figure 6.12 Homeostatic control of osmoregulation.

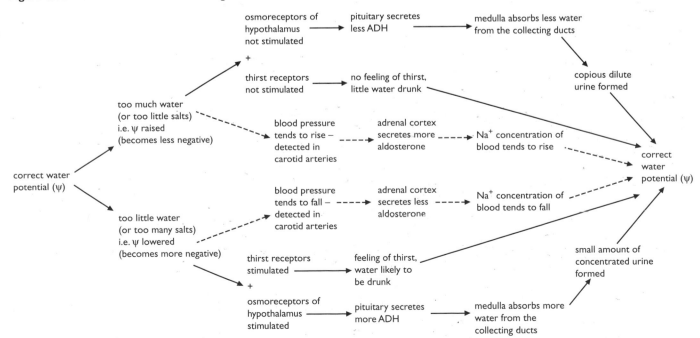

7 Homeostasis

Homeostasis (literally 'staying similar') is the maintenance of a stable internal environment. The effective 'internal environment' of a mammal is the blood circulating around the body, together with the tissue fluid that bathes the cells. In mammals, the components of the internal environment that are closely regulated include:

- body temperature (page 54);
- blood sugar level (page 50);
- water content (page 47);
- ion concentration, e.g. Na$^+$ (page 46);
- pH (page 46);
- blood pressure (page 38).

The liver as an homeostatic organ

The liver consists of thousands of polygonal blocks called lobules, each about 1 mm in diameter. Lobules consist of radiating rows of liver cells called **acini**, separated by blood spaces, the **sinusoids**. Blood is in direct contact with liver cells, but phagocytic cells (Kupffer cells) partly line the sinusoids. The liver has a double blood supply via the hepatic artery (oxygenated blood) and the hepatic portal vein (blood carrying the products of digestion absorbed in the ileum). Blood is drained from the liver by the hepatic vein. Bile, a product of the liver, is carried by the bile duct and, in some mammals, stored in a gall bladder nearby.

> 1 What types of organelle would you expect to see when a liver cell is examined by electronmicroscopy. Why?

Figure 7.1 The liver and its blood supply.

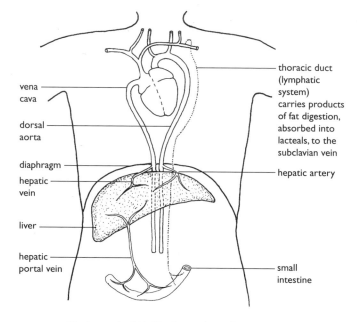

bile duct and gall bladder (humans) not shown

Figure 7.2 The structure of the liver lobules.

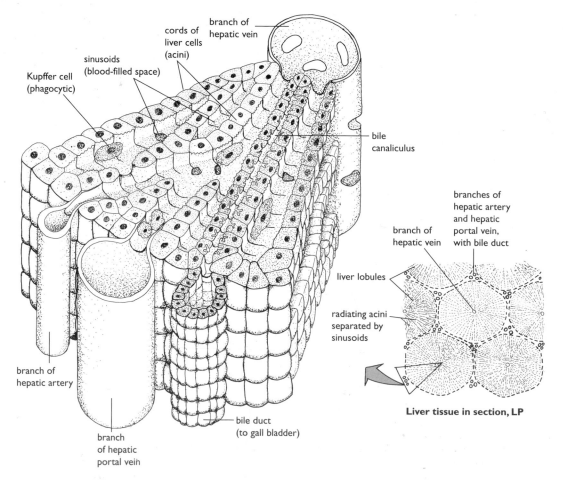

Liver tissue in section, LP

Liver lobule, LP (x 80)

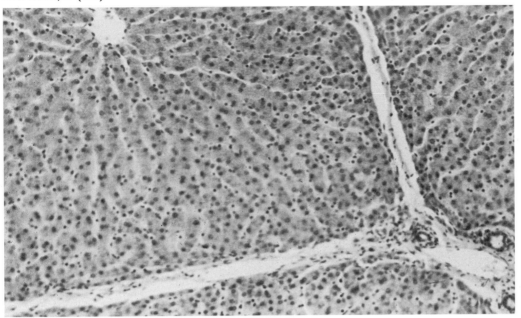

Figure 7.3 Pig liver tissue in section.

Arteriole, interlobular vein and bile duct between lobules, HP (x 250)   **Hepatic vein at centre of lobule, HP (x 250)**

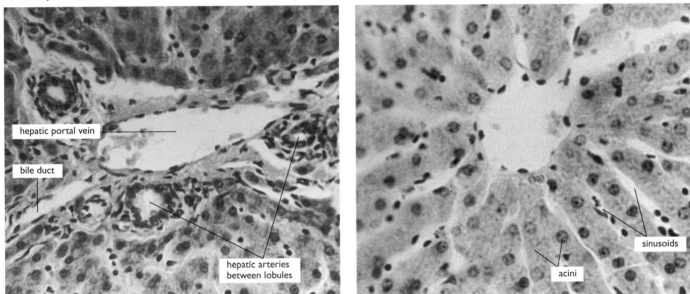

Figure 7.3 Pig liver tissue in section.

I The liver and amino acids – deamination

Proteins function as enzymes, components of membranes, and structural components (e.g. collagen fibres, keratin). They are built from the pool of amino acids maintained in the body. Most proteins are short-lived, and are broken down to contribute to the amino acid pool. About 20 different amino acids are involved, in differing proportions, in protein synthesis.

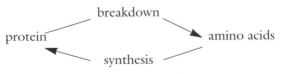

The body cannot store any amino acids that are excess to the amino acid pool: they are deaminated in the liver. The organic acid part of each amino acid is removed and respired, or converted into fat or carbohydrate. The $-NH_2$ group(s) are converted into ammonia and combined with carbon dioxide to form urea:

$$2NH_3 + CO_2 \longrightarrow (NH_2)_2C{=}O + H_2O$$

This deamination prevents the formation of soluble, toxic ammonia in the tissues. Urea is removed from the blood in the kidneys (page 44).

2 The liver and the regulation of blood sugar levels

Tissues receive their glucose supply from the blood circulation. For most tissues it is the principal substrate for respiration. Glucose is quickly absorbed across the cell membrane, and mitochondria contain many of the enzymes required to respire glucose and produce ATP. In addition to fatty acids, glucose is one of the substrates of skeletal muscles; in the brain, glucose is the only 'fuel' molecule absorbed (neurones are unable to store glycogen and they cannot respire lipids).

The normal level of glucose in human blood is about $90\,mg/100\,cm^3$, but the actual concentration varies between $70\,mg$ (when the body has been without food for a prolonged period) and $150\,mg$ (in the hepatic portal vein after a carbohydrate-rich meal has been digested and is being absorbed).

3 In the complete respiration of glucose to carbon dioxide and water, which steps occur in mitochondria and which occur in the cytoplasm?

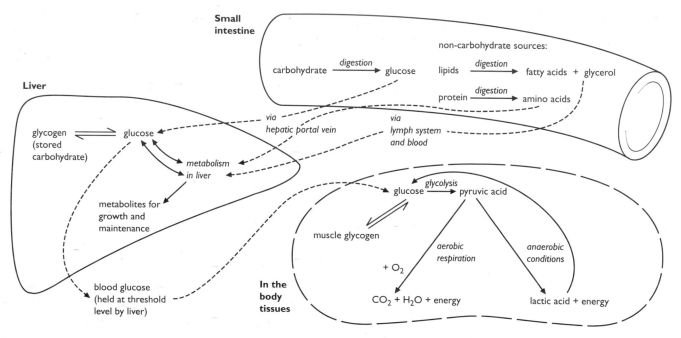

Figure 7.4 How energy in glucose is made available to the body in metabolism.

The pancreas and glucose regulation

The pancreas is important in digestion (page 19). It also contains patches of cells organised into areas called **islets of Langerhans**. These are endocrine glands, that is, they secrete hormones into the blood circulation. Islets of Langerhans contain two types of cell: alpha cells, which produce **glucagon**, and beta cells, which produce **insulin**. Both hormones have a role in glucose regulation, but have antagonistic effects. Glucagon exclusively affects the liver whereas insulin affects both liver and muscle metabolism.

4 What is the difference between an endocrine and an exocrine gland?

Figure 7.5 TS of pancreas showing islet of Langerhans, HP (×400).

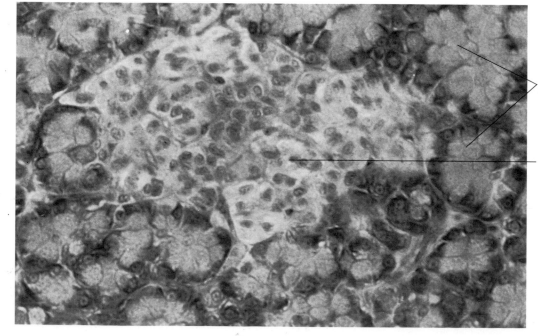

exocrine cells of pancreas

islet of Langerhans

'Self-regulation' of metabolism

Negative feedback is a type of control mechanism commonly involved in homeostasis. It occurs when a departure from the normal (when the blood sugar level falls below (hypoglycemia) or rises above (hyperglycemia) the threshold level) is sensed by a cell, tissue or organ. The corrective response acts to return the variable condition to a normal level.

5 What are the essential components of a physiological feedback mechanism?

The sites of glucose regulation

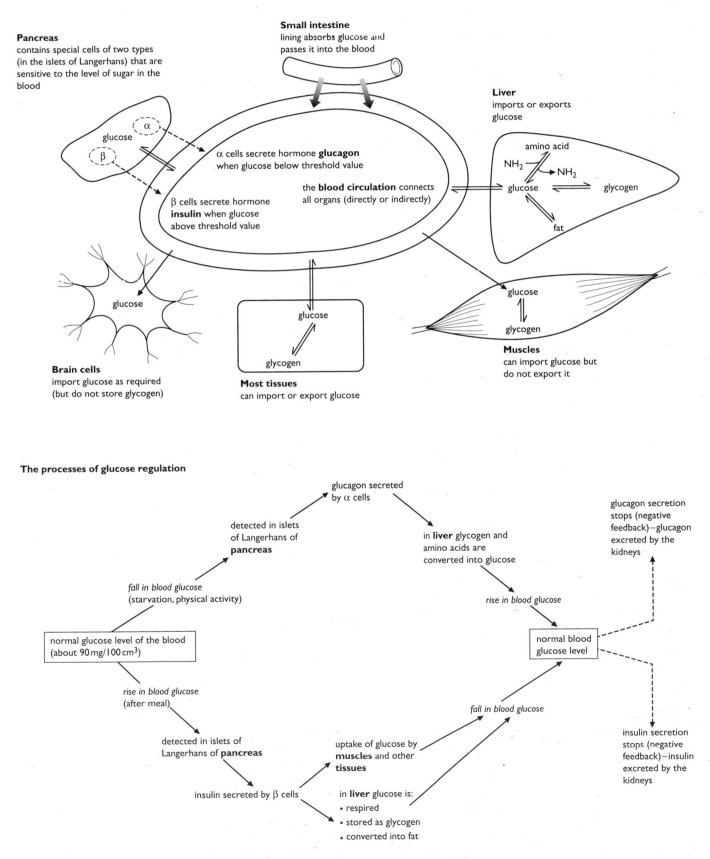

Figure 7.6 The processes and sites of glucose regulation.

3 The liver and bile formation

Bile is a yellow–green alkaline mucous fluid that is formed in the liver. It has key roles in digestion in the duodenum and absorption in the small intestine (page 19). Production of bile is tied-in with the lysis (breakdown) of red cells in the liver lobules.

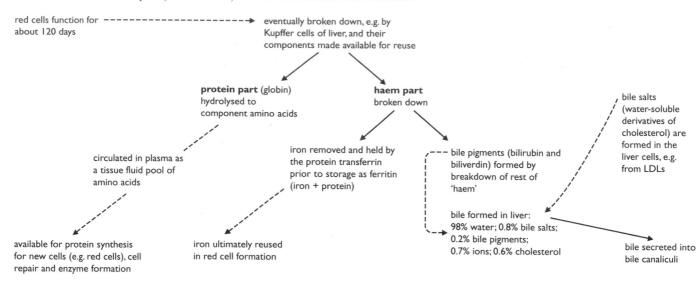

red cells function for about 120 days ----→ eventually broken down, e.g. by Kupffer cells of liver, and their components made available for reuse

protein part (globin) hydrolysed to component amino acids

haem part broken down

circulated in plasma as a tissue fluid pool of amino acids

iron removed and held by the protein transferrin prior to storage as ferritin (iron + protein)

bile pigments (bilirubin and biliverdin) formed by breakdown of rest of 'haem'

bile salts (water-soluble derivatives of cholesterol) are formed in the liver cells, e.g. from LDLs

available for protein synthesis for new cells (e.g. red cells), cell repair and enzyme formation

iron ultimately reused in red cell formation

bile formed in liver: 98% water; 0.8% bile salts; 0.2% bile pigments; 0.7% ions; 0.6% cholesterol

bile secreted into bile canaliculi

Figure 7.7 Breakdown of red cells and the formation of bile. All these aspects of metabolism are carried out in liver cells.

4 The liver and fat metabolism

In the small intestine the digestion products of dietary lipids are absorbed and transported to the liver (Figure 3.17, page 20) with the aid of bile salts. In the liver, fatty acids and glycerol are prepared for metabolism or for storage by the body. Some fatty acids are converted into cholesterol via the respiratory intermediate, acetyl coenzyme A.

Lipids are used as a respiratory substrate to produce energy in muscles; as an energy store in adipose tissue; laid down in connective tissues around the body organs; as cholesterol for the production of steroid hormones; and for the production or repair of cell membranes. They are insoluble in water and have to be transported from the liver in the blood plasma associated with proteins, in components known as very low density lipoproteins (VLDLs), low density lipoproteins (LDLs), or high density lipoproteins (HDLs), named according to the relative proportions of proteins and lipids.

	Protein (%)	Lipid (%)	Particle diameter (nm)
VLDL	10	61	30–90
LDL	27	5	20–25
HDL	50	3	7–10

Figure 7.8 TS of adipose tissue, LP (×400).

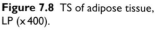

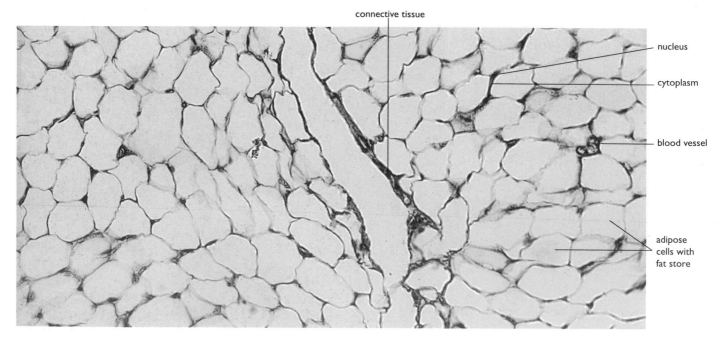

connective tissue

nucleus

cytoplasm

blood vessel

adipose cells with fat store

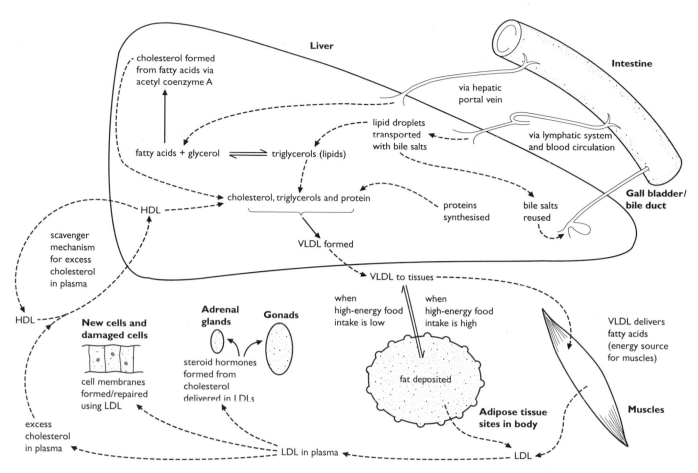

Figure 7.9 A summary of lipid metabolism, showing the key role of the liver.

5 Other roles of the liver

- **Plasma proteins:** the liver is the site of plasma protein manufacture, including those associated with blood clotting (page 35) and albumin, which helps buffer pH change, and exerts an osmotic effect in tissue fluid production and reabsorption.
- **Storage of vitamins:** lipid-soluble vitamins (A, D, E and K) are stored here, and so too is vitamin B12. Vitamin A is actually manufactured here, from the precursor molecule carotene.
- **Detoxification:** the liver inactivates and prepares for excretion hormones and substances absorbed into the body, such as drugs.

Alcohol and liver function

Cirrhosis is a disease of the liver where cells are destroyed and replaced by fibrous collagen tissue, and the blood supply to liver cells is impaired. Production of essential proteins such as albumin and blood-clotting proteins is interrupted. Failure to remove nitrogenous waste products from the blood leads to brain damage.

Alcohol (ethanol) abuse is a common cause of cirrhosis in patients in the developed world. Ethanol is converted into ethanal (acetaldehyde) in the liver. Ethanal is toxic and responsible for many of the features of cirrhosis. The risk of developing cirrhosis is related to the amount of ethanol consumed daily, but there is evidence that genetic factors may predispose some people to the disease.

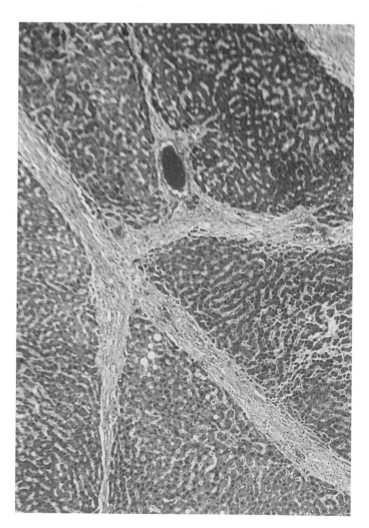

Figure 7.10 Liver tissue harmed by alcohol. The normal architecture of liver lobules is distorted by the formation of fibrous tissue in place of dead liver cells.

Temperature regulation

Animals that can regulate their own body temperature have a better chance of survival in environments where external temperatures fluctuate. Mammals (and birds) maintain a more or less constant body temperature, irrespective of changes in their environment, by generating heat within the body and controlling heat loss. They are called **'endothermic'**.

Mammals may **obtain heat** from:

- **external sources**, directly from solar radiation, or indirectly from radiation, convection and conduction from the environment (itself warmed by solar radiation), or from periods of close contact with other animals;
- **internal sources**, as a by-product of metabolic reactions such as cellular respiration, and from muscular contractions.

Animals may **lose heat** by radiation, convection and conduction from the body surface, or by evaporation of water. The body's regulation of heat loss mainly occurs through the external skin, although in many furry animals the interior of their snouts (nose and mouth) is also an important surface for heat loss.

6 At what times or under what conditions does the human body temperature typically vary from the norm (approximately 36.8 °C) during a normal 24-hour period?

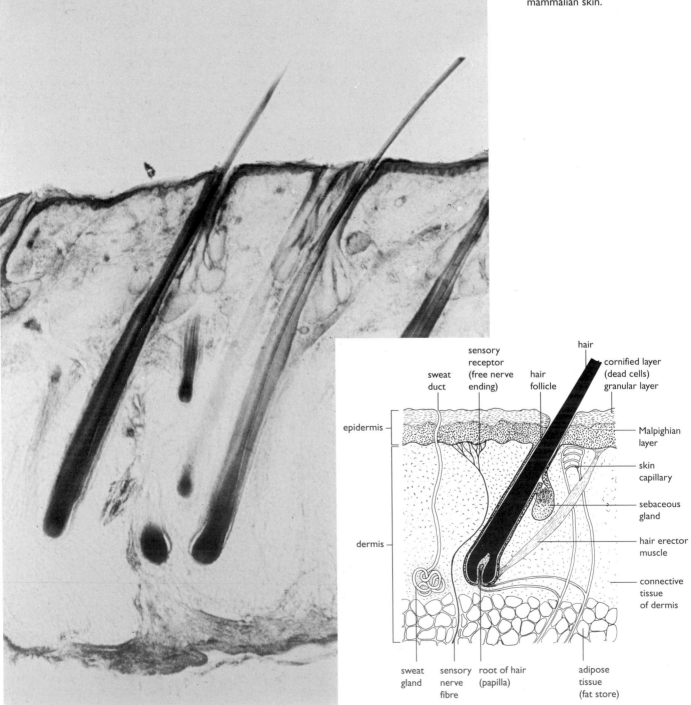

Figure 7.11 The structure of mammalian skin.

Mechanisms to combat cold

- Hairs are raised, increasing the insulating layer of stationary air at the body surface.
- The blood flow between the body 'core' and the skin is decreased by constriction of the skin arterioles.
- Secretion of sweat is stopped, reducing loss of heat by evaporation.
- Muscle tone is raised, tending to generate additional heat. In extreme cold, shivering occurs, generating heat in muscles that may need to be active.
- Mammals may huddle together if the cold is severe.

Fat stored below the skin is an energy reserve. It helps to reduce heat loss only if minimally supplied with blood capillaries.

Mechanisms to combat overheating

- Hairs are lowered, reducing the insulating layer of stationary air at the body surface.
- The blood flow between the body 'core' and the skin is increased by dilation of the skin arterioles.
- Secretion of sweat occurs, enhancing heat loss by evaporation.
- Behavioural mechanisms are employed, such as moving into the shade, decreasing physical activity, immersion in water or licking of fur, and the practice of panting in animals with significant fur and limited sweat glands.

Figure 7.12 The homeostatic mechanism of temperature control in mammals.

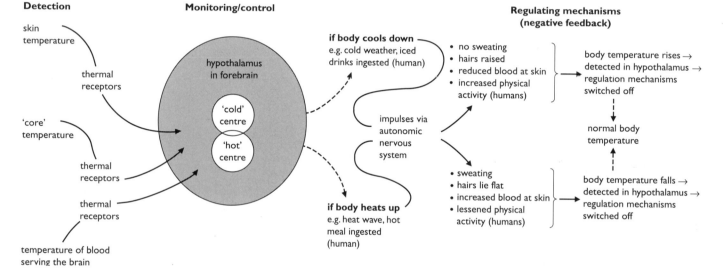

Hibernation in mammals

When temperatures are low and food (e.g. insects) is unavailable, certain species of small mammals may become torpid. The body cools down to approximately the same temperature as the surroundings. Bats do this both in winter and in cold periods in summer. Protected sites are chosen, and the animals often huddle together. They are still vulnerable to predators.

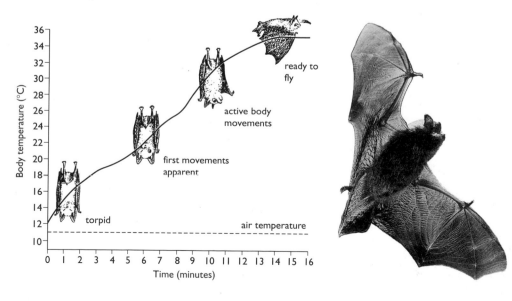

Figure 7.13 The pipistrelle bat (*Pipistrellus pipistrellus*) active and torpid.

Survival in arid habitats

In desert habitats rain is rare and unpredictable. There is little vegetation cover, so soils are subject to extremes of heat by day and cold by night. With few obstacles to slow down the air moving over the land, and little root growth or surface moisture to bind soil together, wind tends to shape the landscape of deserts in a way water does in most other habitats. Yet some mammals (and other animals) can survive desert conditions.

1 **Camels** are members of the 'even-toed, hoofed mammals'. Today the two types of camel, the dromedary (one-humped) and the Bactrian (two-humped), are largely reduced to being domestic animals. The dromedary's single hump is better adapted to hotter climates because (i) the single hump has a smaller surface area than a double hump, and (ii) it is covered by elastic skin, which contracts when the fat store is used up, rather than 'flopping over' as do the humps of the Bactrian. These two types of camel readily interbreed and produce fertile progeny.

Figure 7.14 *Camelus bactrianus* (Bactrian camel).

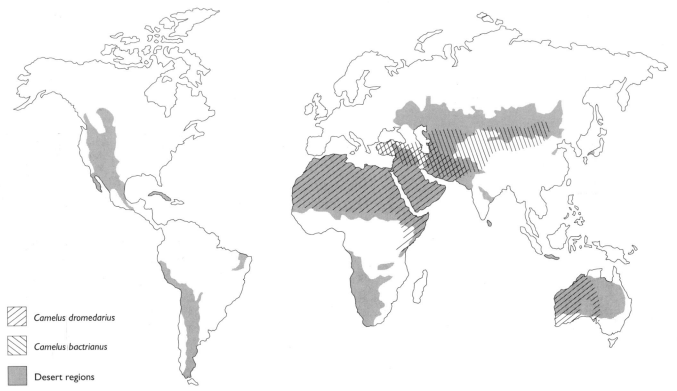

Camelus dromedarius

Camelus bactrianus

Desert regions

Figure 7.15 Distribution of desert regions and of areas where the majority of camels occur.

- **Water conservation by heat storage:** Figure 7.16 shows how the body temperature of a camel fluctuates when it has access to unlimited water (variation of 2 °C) and when it experiences dehydration (variation of up to 7 °C). Water that would otherwise evaporate is saved: for a 500 kg camel, a rise of 7 °C is equivalent to a saving of 5 litres of water. In the cool night period that follows, the heat stored in the body mass by day is lost by conduction and radiation.

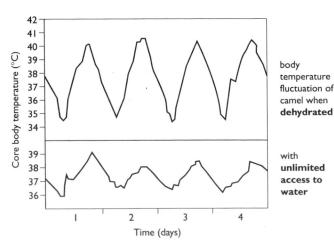

body temperature fluctuation of camel when **dehydrated**

with **unlimited access to water**

Time (days)

Figure 7.16 A study of core temperature fluctuations in a camel.

- **Water vapour saving during breathing:** the lining of the nostrils dries out and becomes covered with dried mucus. When air is breathed out from the lungs, the water vapour present is largely retained by the hygroscopic properties of the mucus, but evaporates in the dry air of the next inhalation. So water vapour is shunted between lungs and nostrils, rather than lost from the body.
- **Heat insulation:** a thick fur coat with a high insulation value reduces heat gain from the environment.
- **Toleration of water depletion:** a loss of almost 25% of body weight may occur in the desert, going without drinking for 6–8 days at a time. By contrast, humans can endure a loss of only 10–12% of body weight. But, when water is available the camel takes in up to one-third of its body weight in water immediately. However, there is no water storage by this animal. Camels are the only mammals that do not have circular erythrocytes. Instead, they are a robust, ovoid shape that is maintained through normal dehydration periods, which allows the maintenance of a smooth flow of blood through the narrowest capillaries.
- **Concentrated urine:** the kidney tubules have relatively long loops of Henle (page 45), and very concentrated urine is formed. Therefore, the drinking of 'brackish' waters with quite high salt concentrations is possible.
- **The digestive system:** although they are not specifically ruminants (but closely related to them), camels do 'ruminate' (page 23). There is a forestomach (equivalent to the rumen + reticulum), and a tubular stomach (equivalent to the abomasum). The bolus of food is brought back to the mouth, for re-chewing, before being passed to the tubular stomach and the rest of the alimentary canal. The available vegetation is thus fully exploited to obtain an adequate diet.
- **Adaptations of feet, nostrils and mouth:** the non-hoofed, padded feet of the camel are ideal for movement on desert surfaces. The nostrils are slit-like and guarded by dense hairs. The eyes are shielded by long, dense eye lashes and an additional eye lid. The lips are split and highly mobile, allowing their use for selective browsing on foliage protected by thorns.

7 What are the implications for their classification that camels interbreed and produce fertile offspring?

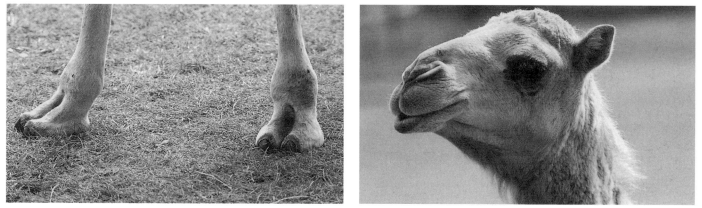

Figure 7.17 Non-hoofed feet and head (nostrils and mouth) of a dromedary camel (*Camelus dromedarius*).

2 The **kangaroo rat** (*Dipodomys*) lives in hot, dry deserts in southwest America, where water is not usually available. These animals obtain their only metabolic water from a diet of dry seeds. Extremely concentrated urine is produced, and no sweating occurs. Survival is possible by remaining well below ground during the heat of the day. Respiratory moisture is trapped in the nostrils, as in the desiccated camel (see above).

Figure 7.18 Kangaroo rat (*Dipodomys* sp.).

Water gain		Water loss	
Food (dry seed) in equilibrium with air at 20% r.h. contains	6.0 g	Breathing	43.9 g
		Discharge of 3.17 g of urea requires	13.5 g
Metabolic water obtained by oxidation of food	54.0 g	Defecation	2.6 g
Total	60.0 g		60.0 g

Water balance per 100 g of food consumed in a 4-week period.
Air temperature = 25 °C; relative humidity = 20%.

8 Sensitivity, coordination and control

Response to stimuli, external and internal, and coordination of all the body's systems is achieved by two distinct but inter-related mechanisms. In the **nervous system**, information between different parts of the body passes as electrochemical impulses through specialised nerve cells. In the **endocrine system**, chemical hormones, produced in ductless glands (endocrine glands), are carried all over the body in the blood but trigger changes in specific systems. The brain has an overarching role in the integration and control of the nervous and endocrine systems.

The nervous system

The nervous system consists of specialised cells, called neurones. A **neurone** has a cell body containing the nucleus and the bulk of the cytoplasm, a fine cytoplasmic fibre (dendron) or several such fibres (dendrites), which bring impulses towards the cell body, and a single fibre (axon) taking impulses away from the cell body. Associated with the neurones are **sense organs** (called sensory receptors) that detect stimuli and convert them into impulses, and **effectors**, which are muscles or glands that respond when stimulated by impulses. Sense organs and effectors are connected by neurones arranged in pathways called **reflex arcs**.

Relay neurones are structurally simple in comparison with **sensory** neurones and **motor** neurones (see Figure 8.3). Relay neurones are so named because of their positions in reflex arcs in relation to sense organs and effector organs (see Figure 8.5).

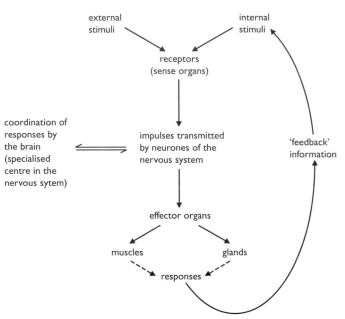

Figure 8.1 The steps to coordination and control by the nervous system.

Figure 8.2 A relay neurone.

Relay neurones occur within the central nervous system (brain and spinal cord). They may also be referred to as intermediate or internuncial neurones

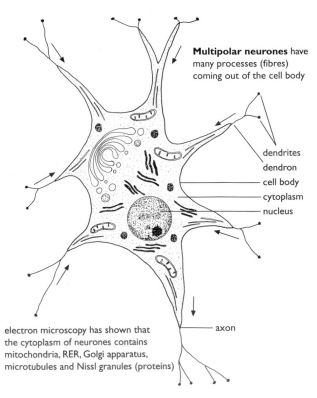

Multipolar neurones have many processes (fibres) coming out of the cell body

- dendrites
- dendron
- cell body
- cytoplasm
- nucleus

electron microscopy has shown that the cytoplasm of neurones contains mitochondria, RER, Golgi apparatus, microtubules and Nissl granules (proteins)

axon

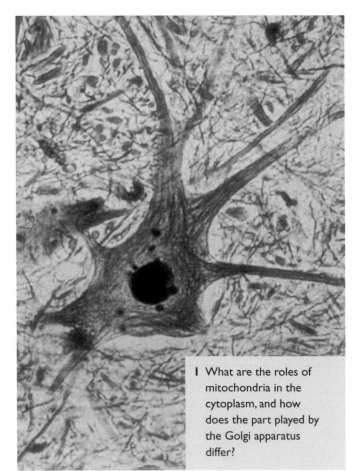

1 What are the roles of mitochondria in the cytoplasm, and how does the part played by the Golgi apparatus differ?

Dendrons and axons are extremely thin fibres that transmit impulses at high speeds, typically between 30 and 120 m/s in mammals. These tiny fibres are insulated and protected from mechanical damage and electrical interference by special myelin sheaths, formed from supporting cells called Schwann cells. Junctions in the sheaths are called nodes of Ranvier. These help in impulse transmission.

Figure 8.3 Sensory and motor neurones and their myelin sheaths formed from Schwann cells.

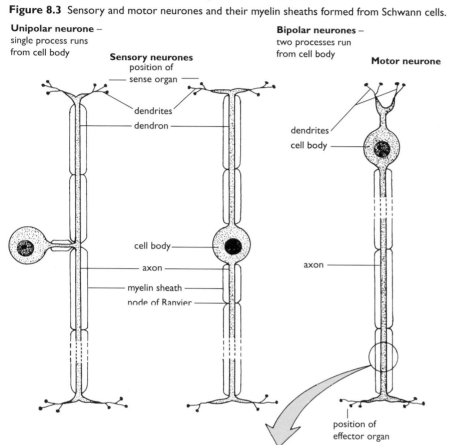

Unipolar neurone – single process runs from cell body

Bipolar neurones – two processes run from cell body

Sensory neurones
position of sense organ
dendrites
dendron
cell body
axon
myelin sheath
node of Ranvier

Motor neurone
dendrites
cell body
axon
position of effector organ

myelination of nerve fibre by the wrapping of many turns of the Schwann cell membrane, forming a myelin sheath

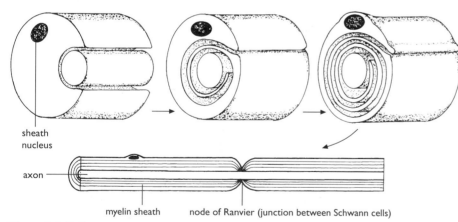

sheath nucleus
axon
myelin sheath
node of Ranvier (junction between Schwann cells)

Figure 8.4 LS peripheral nerve, HP.

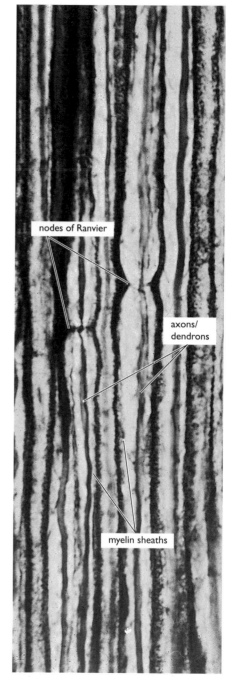

nodes of Ranvier
axons/ dendrons
myelin sheaths

Figure 8.5 Reflex arcs.

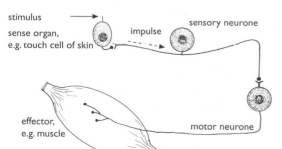

Reflex arc of sensory and motor neurone only: (monosynaptic reflex)

stimulus
sense organ, e.g. touch cell of skin
impulse
sensory neurone
effector, e.g. muscle
motor neurone

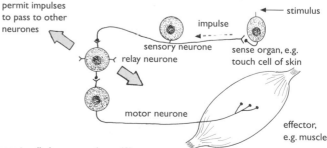

Reflex arc with a relay neurone included: (polysynaptic reflex)

relay neurones permit impulses to pass to other neurones
stimulus
impulse
sensory neurone
sense organ, e.g. touch cell of skin
relay neurone
motor neurone
effector, e.g. muscle

the working junction between neurones is called a synapse (page 62)

Investigation of the impulse

In some of the nonvertebrates, including the squid and the earthworm, the nervous system contains some axons known as 'giant fibres', which can reach a millimetre in diameter. These are associated with a rapid transmission of impulses. Giant fibres can be removed from an anaesthetised animal and are large enough for a microelectrode to be inserted into isolated lengths of axon.

The **discovery of the resting potential** came from the use of an external and internal microelectrode on an isolated axon (a giant fibre), the amplification of the signal and its display on a cathode ray oscilloscope (CRO). In the 'resting' nerve there was a potential difference between the inside and the outside of the fibre of approximately −70 mV. This potential difference is due to two processes:

- vigorous active transport of K^+ ions in and Na^+ ions out of the axon;
- facilitated diffusion that allows much more K^+ out than Na^+ back in.

2 What are the differences between active transport and facilitated diffusion of ions?

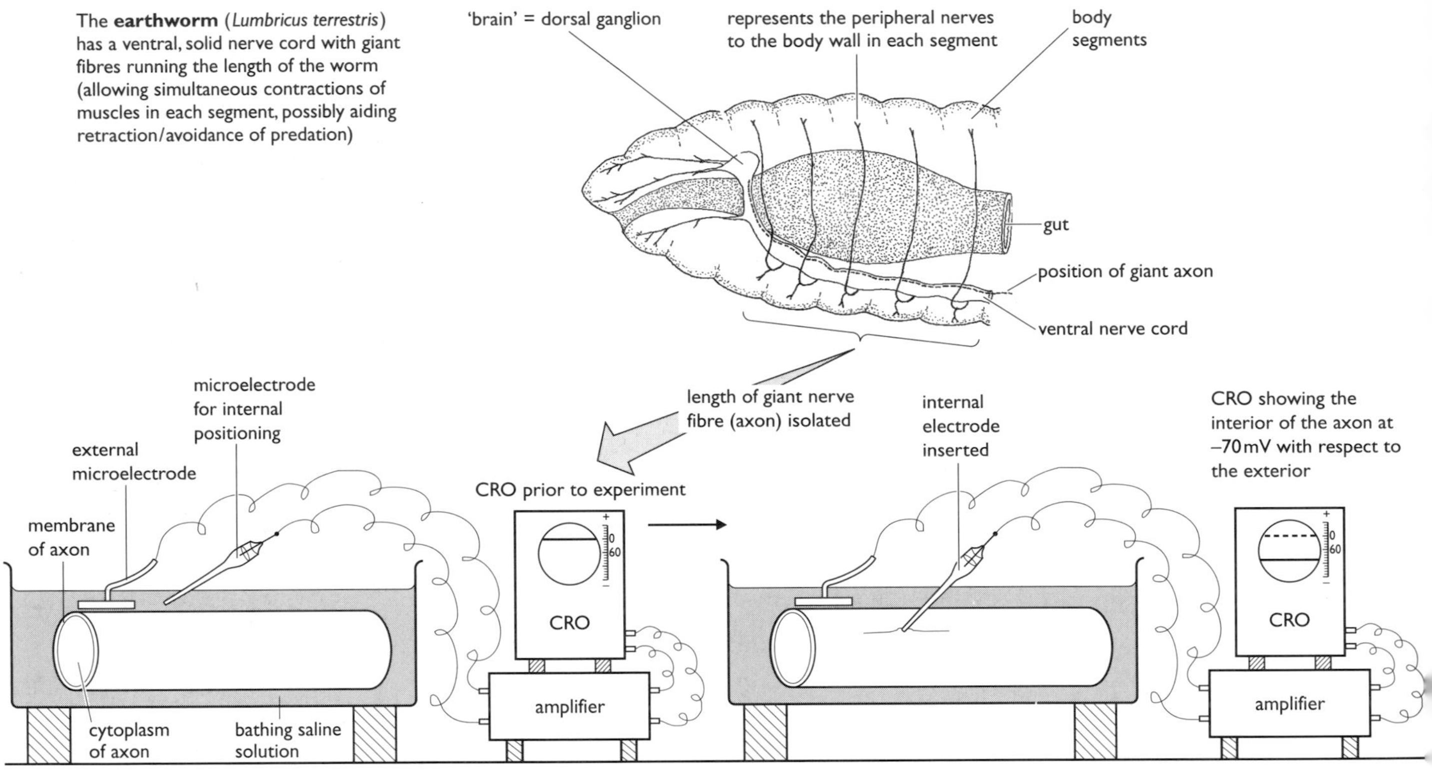
The **earthworm** (*Lumbricus terrestris*) has a ventral, solid nerve cord with giant fibres running the length of the worm (allowing simultaneous contractions of muscles in each segment, possibly aiding retraction/avoidance of predation)
'brain' = dorsal ganglion
represents the peripheral nerves to the body wall in each segment
body segments
gut
position of giant axon
ventral nerve cord
length of giant nerve fibre (axon) isolated
microelectrode for internal positioning
external microelectrode
membrane of axon
cytoplasm of axon
bathing saline solution
CRO prior to experiment
CRO
amplifier
internal electrode inserted
CRO showing the interior of the axon at −70 mV with respect to the exterior

Figure 8.6 Investigation of resting potential.

Figure 8.7 Formation of the resting potential.

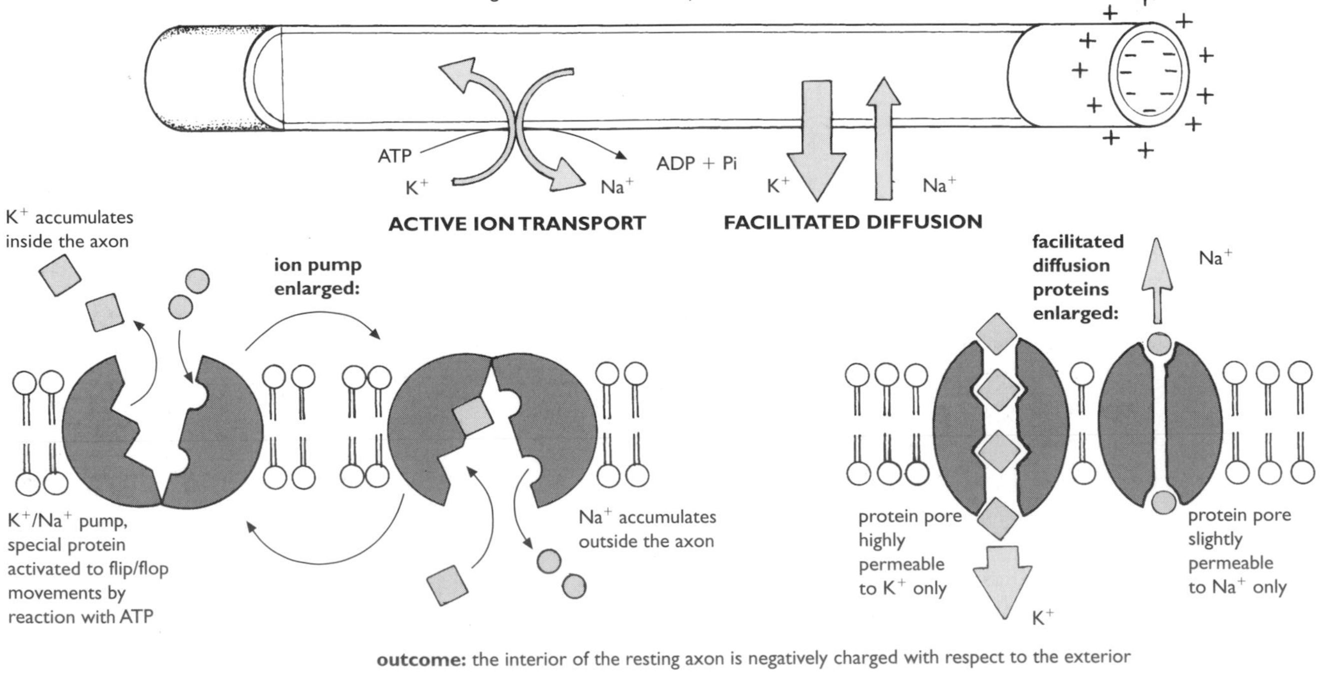
length of axon shown cut open
ATP
ADP + Pi
K+
Na+
ACTIVE ION TRANSPORT
FACILITATED DIFFUSION
K+ accumulates inside the axon
ion pump enlarged:
K+/Na+ pump, special protein activated to flip/flop movements by reaction with ATP
Na+ accumulates outside the axon
facilitated diffusion proteins enlarged:
protein pore highly permeable to K+ only
protein pore slightly permeable to Na+ only
outcome: the interior of the resting axon is negatively charged with respect to the exterior

The **impulse (action potential)** is a temporary and local reversal of the resting potential, which occurs when the axon is stimulated. During an action potential, the membrane potential falls until the inside of the membrane becomes positively charged with respect to the outside, a change from -70 mV to $+40$ mV. The membrane is then described as 'depolarised'. Depolarisation is a temporary condition. The action potential runs the length of the neurone fibre, but at any one point it exists for only 2 milliseconds ($\frac{2}{1000}$ of a second) before the resting potential is re-established. So, action potential transmission is exceedingly quick, and occurs because of pores (ion channels) in the axon membrane that open and close momentarily, regulating the flow of ions.

3 Why is an axon unable to conduct an impulse immediately after an action potential has been conducted?

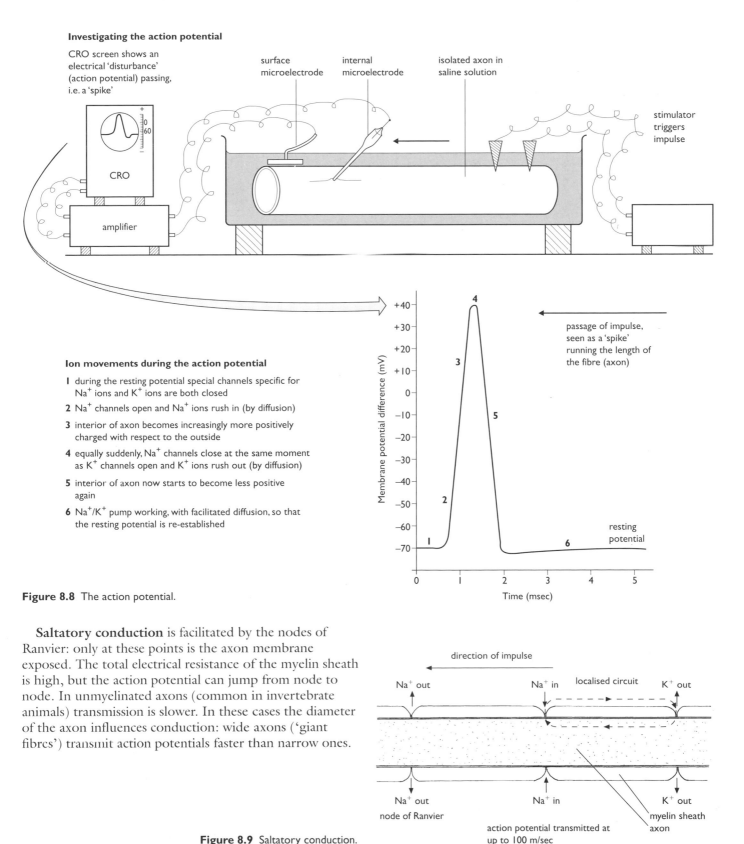

Investigating the action potential

CRO screen shows an electrical 'disturbance' (action potential) passing, i.e. a 'spike'

surface microelectrode

internal microelectrode

isolated axon in saline solution

stimulator triggers impulse

CRO

amplifier

passage of impulse, seen as a 'spike' running the length of the fibre (axon)

Ion movements during the action potential

1 during the resting potential special channels specific for Na$^+$ ions and K$^+$ ions are both closed

2 Na$^+$ channels open and Na$^+$ ions rush in (by diffusion)

3 interior of axon becomes increasingly more positively charged with respect to the outside

4 equally suddenly, Na$^+$ channels close at the same moment as K$^+$ channels open and K$^+$ ions rush out (by diffusion)

5 interior of axon now starts to become less positive again

6 Na$^+$/K$^+$ pump working, with facilitated diffusion, so that the resting potential is re-established

Membrane potential difference (mV)

resting potential

Time (msec)

Figure 8.8 The action potential.

Saltatory conduction is facilitated by the nodes of Ranvier: only at these points is the axon membrane exposed. The total electrical resistance of the myelin sheath is high, but the action potential can jump from node to node. In unmyelinated axons (common in invertebrate animals) transmission is slower. In these cases the diameter of the axon influences conduction: wide axons ('giant fibres') transmit action potentials faster than narrow ones.

direction of impulse

Na$^+$ out   Na$^+$ in   localised circuit   K$^+$ out

Na$^+$ out   Na$^+$ in   K$^+$ out

node of Ranvier

myelin sheath
axon

action potential transmitted at up to 100 m/sec

Figure 8.9 Saltatory conduction.

The synapse

An action potential is transmitted from one neurone to another at a structure called a synapse. At this point there is a tiny gap between the meeting neurones about 20 nm wide, called the **synaptic cleft**. Action potentials do not cross this gap; transmission at the synapse is by a specific chemical, known as a **neurotransmitter substance** (also see page 67). Release of this transmitter at the presynaptic membrane is triggered by the arrival of the action potential 'spike'. The transmitter diffuses across the gap in about 0.5 ms, and *may* initiate an action potential in the postsynaptic neurone.

Transmitter substances are relatively small, diffusable molecules, for example acetylcholine, noradrenaline. In the brain, glutamic acid and dopamine are the commonly occurring transmitters. Some 40 different compounds are believed to function as neurotransmitters within our nervous system. Transmitter substances are manufactured and packaged in the Golgi apparatus of neurones.

Figure 8.10 A synapse seen in section.

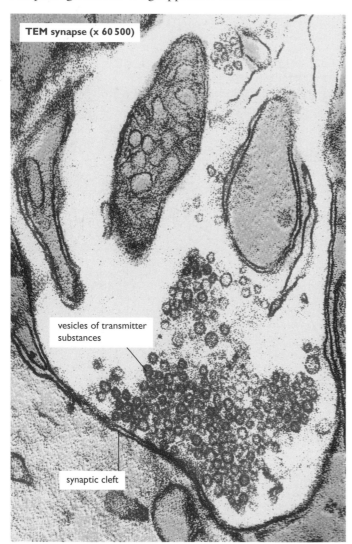

TEM synapse (x 60 500)

vesicles of transmitter substances

synaptic cleft

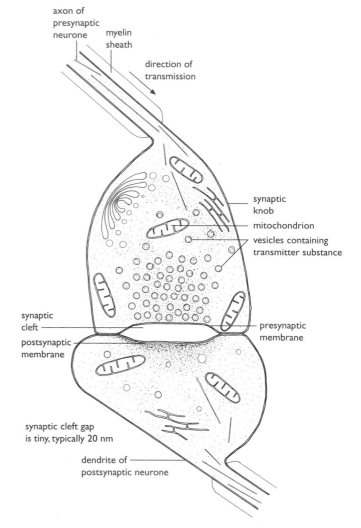

axon of presynaptic neurone

myelin sheath

direction of transmission

synaptic knob
mitochondrion
vesicles containing transmitter substance

synaptic cleft
postsynaptic membrane

presynaptic membrane

synaptic cleft gap is tiny, typically 20 nm

dendrite of postsynaptic neurone

Drugs can alter activity at synapses. Some substances amplify the process of synaptic transmission (**agonistic drugs**):

- **amphetamines** cause increased release of noradrenaline leading to enhanced activation of neurones in the brain;
- **cocaine** causes noradrenaline to persist in the synaptic cleft, with similar consequences to those of amphetamines;
- **nicotine**, of similar structure to the active part of the acetylcholine molecule, fits on to acetylcholine acceptor molecules but is not broken in the same way. It remains attached to the receptors, thus prolonging the effects.

Other drugs can inhibit synaptic transmission (**antagonistic drugs**):

- **atropine** inhibits acetylcholine, preventing an action potential from being generated;
- **curare** similarly blocks nerve/muscle junctions (page 75).

4 What are the principal effects of the 'socially acceptable' drug ethanol upon the functioning of the nervous system?

In the process of **chemical transmission** at the synapse, the transmitter binds to the receptor site and typically, causes the opening of Na^+ channels. In the case of an **excitatory synapse**, an action potential is generated in the postsynaptic neurone, as shown in Figure 8.11. However, in some synapses the reverse happens. That is, arrival of the transmitter substance causes the entry of Cl^- ions and the release of K^+ ions, so that the interior of the postsynaptic neurone becomes even more negative. Passage of an action potential is thus inhibited. This is called an **inhibitory synapse**.

Figure 8.11 Chemical transmission at the synapse.

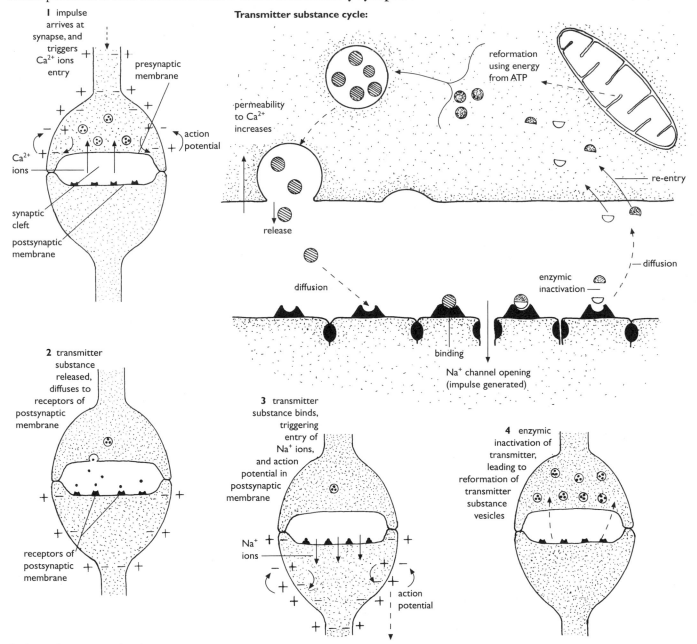

Roles of synapses in the nervous system

- The synapse conveys information between neurones in **one direction only**.
- A postsynaptic neurone may have many synapses with different presynaptic neurones. At each synapse the action potential may lead to **excitation or inhibition** of an action potential. Thus action potentials arriving at the synapse may tend to reinforce (summation) or cancel each other. The neurone + synapse system allows the reception of impulses from different sources and allows an appropriate, **subtle pattern of response**.
- Synapses filter out any infrequent and low-level stimuli, thereby **removing all 'background noise'** from the body's communication system.
- The synapse **protects the response system from overstimulation** by becoming fatigued. Continuous transmission of action potentials eventually exhausts the immediate supply of transmitters (habituation).
- The presence of millions of neurones with vast numbers of synapses in the brain may be the physical basis of our ability to **learn and memorise**.

The sensory system

Sense organs contain cells that are able to respond to stimuli by the production of an action potential, which is then conducted to other parts of the nervous system. Most sensory cells are especially sensitive to only one type of stimulation – heat, light, touch or chemicals, for instance. Sense organs may occur as individual cells, or as small groups of cells, or merely as specialised nerve endings, for example touch/pressure receptors in the skin; others take the form of a large number of highly specialised cells contained within a specialised organ, for example the eye with its light-sensitive retina (see Figure 8.18, page 67).

5 What is the role and value to the animal of the presence of pressure receptors deep in body tissues, such as those observed around joints?

Pacinian corpuscles occur deep in the skin all over the body, and in the capsules around joints

nerve ending (dendrite)

layers of collagen fibres and flattened cells

dendron of sensory neurone (myelinated)

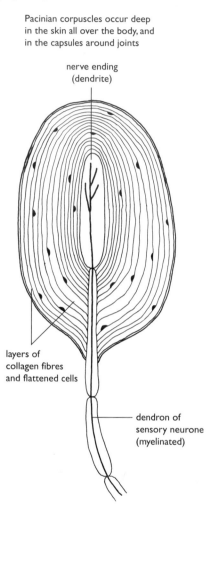

Pacinian corpuscle in cross-section (x 450)

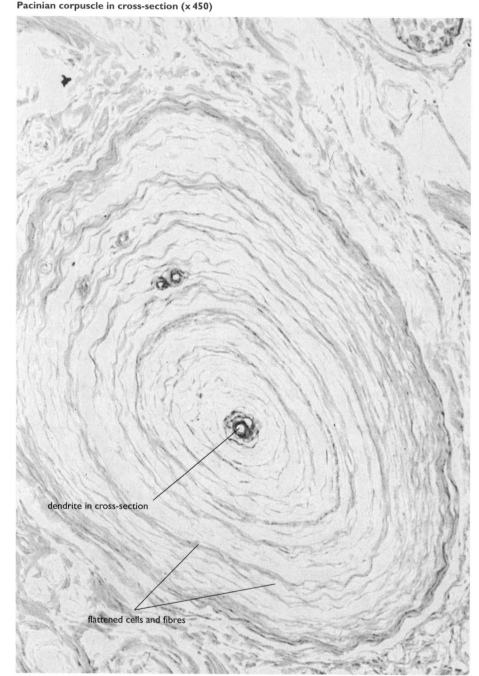

dendrite in cross-section

flattened cells and fibres

Figure 8.12 Pacinian corpuscle, a touch/pressure receptor.

The **generation of an action potential** by a Pacinian corpuscle occurs when localised, strong pressure is applied to our skin, temporarily deforming the layers of collagen within the capsule. As a consequence, the permeability of the nerve ending in the corpuscle to Na^+ ions is temporarily changed: Na^+ ions flow in, and the interior starts to become less negative. At this early stage the depolarisation is called a **generator potential**. The stronger the stimulus, the greater the generator potential. If the generator potential reaches (or exceeds) a pre-set threshold of stimulation it then causes an action potential in the axon.

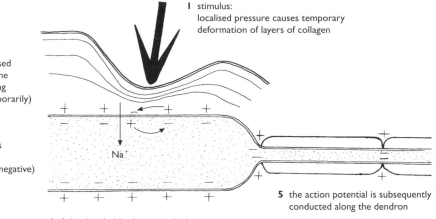

1 stimulus:
localised pressure causes temporary deformation of layers of collagen

2 deformation results in increased permeability to Na$^+$ ions in the membrane of the nerve ending (the Na$^+$ channels open temporarily)

3 a 'generator potential' is created by this influx of Na$^+$ ions into the interior of the nerve ending (which becomes less negative)

Na$^+$

5 the action potential is subsequently conducted along the dendron

4 if the threshold value is reached, an action potential is triggered in the dendron (if not, the resting potential is quickly restored)

The **strength of the stimulus determines the frequency of the action potential** generated by the sense organ. The stronger the stimulus, the more frequently action potentials flow, at least initially. But if the stimulus is continuously maintained on the sense cell at a high level, the permeability of the membrane to Na$^+$ ions changes. The frequency of action potentials slows and then stops; the organ is said to have **adapted**. Different sense organs 'adapt' to varying degrees. Fine-touch receptors in the skin adapt quickly (e.g. after dressing we cease to notice the touch of clothing on skin). Pain receptors all over our bodies, and stretch receptors in our muscles, hardly adapt at all.

The **senses of taste and smell** are detected in the mouth and nose. Many mammals (e.g. the dog) have an extremely well-developed sense of smell, which may play a vital part in their selection of food, detection of danger, establishment of territory and in the finding of a mate. A human's sense of smell is only modestly developed by comparison.

Taste is detected by special taste receptor cells located on the tongue and on the roof of the mouth. Taste receptors are grouped into taste buds.

Smell is detected in the nasal passage. Inhaled vapours stimulate olfactory cells in the roof of the nasal cavity. Several thousands of different odours can be detected by a relatively small number of different types of detector cells, and at very low concentrations. The flavours of our foods are largely detected by our sense of smell.

Figure 8.13 From stimulus to action potential in a Pacinian corpuscle.

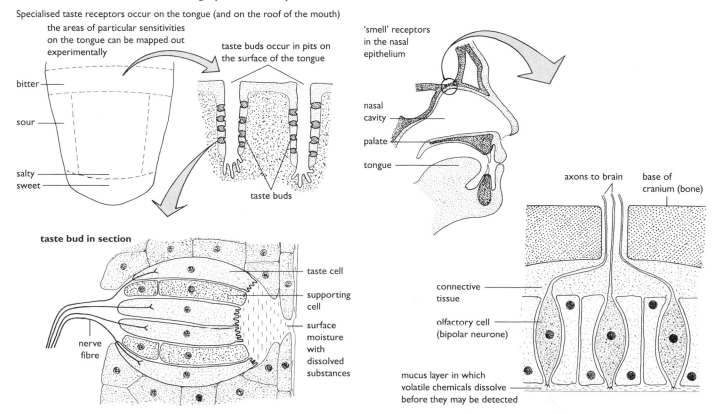

Specialised taste receptors occur on the tongue (and on the roof of the mouth)

the areas of particular sensitivities on the tongue can be mapped out experimentally

taste buds occur in pits on the surface of the tongue

bitter
sour
salty
sweet

taste buds

'smell' receptors in the nasal epithelium

nasal cavity
palate
tongue

axons to brain   base of cranium (bone)

taste bud in section

taste cell
supporting cell
surface moisture with dissolved substances

nerve fibre

connective tissue

olfactory cell (bipolar neurone)

mucus layer in which volatile chemicals dissolve before they may be detected

Figure 8.14 The sensation of 'taste', largely detected on the tongue.

Figure 8.15 The sensation of 'smell', detected in the nasal epithelium.

The sense of sight

The eyes supply information from which the brain deduces the size, shape, movement and (sometimes) the colour of objects in the environment, and about the direction and intensity of light. When both eyes are directed forward, the visual fields of the eyes overlap and the brain resolves the information from the two retinas into a three-dimensional image.

The eye is essentially a three-layered structure. The outer opaque **sclera** contains and protects the eye, but at the front this layer becomes the transparent **cornea**. The middle layer, the **choroid**, is black, and prevents reflection of light within the eyeball. It contains a capillary network. Towards the front of the eye the choroid layer forms the ciliary body and the iris. The inner layer is the **retina**, which consists of neurones and photoreceptor cells. The eyes are connected to the brain by the optic nerves, and the eyeballs are moved in their sockets by rectus muscles.

Figure 8.16 The eye in section.

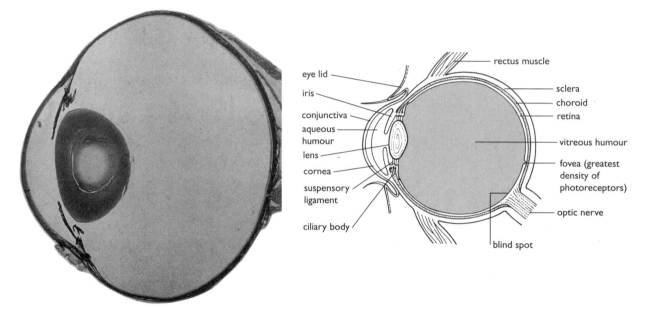

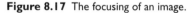

In the focusing process (**accommodation**), light rays from an object enter the eyes and are bent (refracted), mainly by the cornea, to form an image on the retina. However, the degree of refraction by the cornea cannot be varied. The function of the lens is to adjust the degree of refraction to produce a sharp image on the retina. The elastic lens does this by changing shape.

Figure 8.17 The focusing of an image.

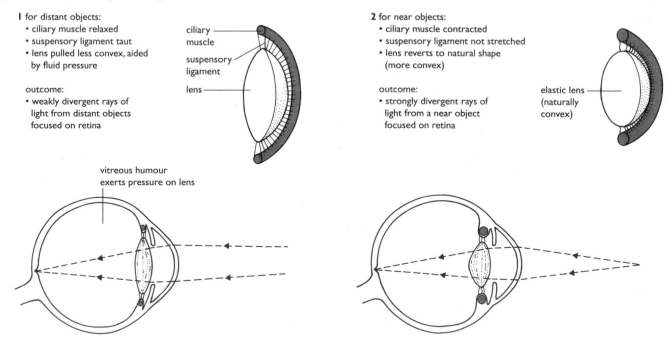

The **retina** contains photoreceptor cells, which are sensitive to electromagnetic radiation in the wavelength range 380–760 nm. Light falling on the retina causes reversible structural change to photopigments in these photoreceptor cells, and the product causes a generator potential (page 64) to arise. As a result, an action potential may be generated in the adjacent ganglion cell, and propagated to the brain along neurone fibres. The photoreceptor cells of the retina are of two types, known as **rods and cones**. The human retina contains about 7 million cones and 10–20 million rods. Cone cells allow sharpness of vision and colour vision.

Figure 8.18 The structure of the retina.

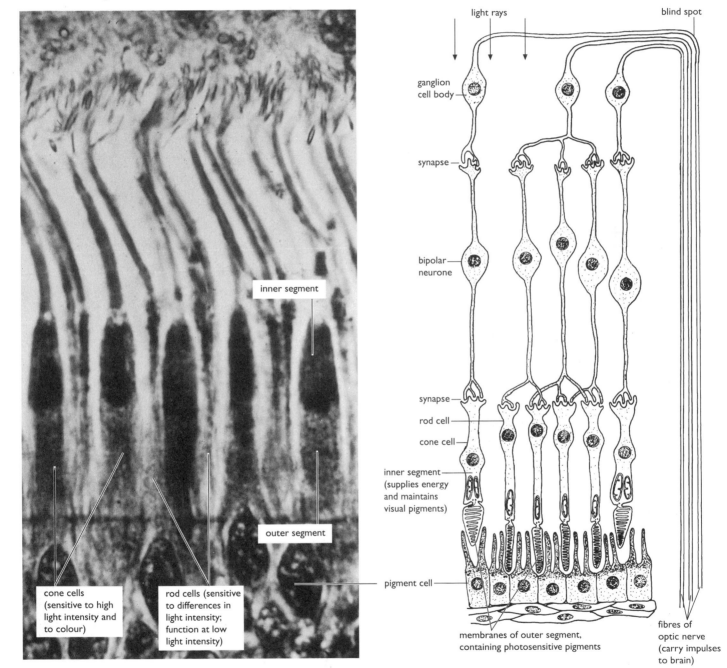

cone cells (sensitive to high light intensity and to colour)

rod cells (sensitive to differences in light intensity; function at low light intensity)

inner segment

outer segment

light rays

blind spot

ganglion cell body

synapse

bipolar neurone

synapse

rod cell

cone cell

inner segment (supplies energy and maintains visual pigments)

pigment cell

membranes of outer segment, containing photosensitive pigments

fibres of optic nerve (carry impulses to brain)

Reversible changes to the photopigments (light-triggered) alter the permeability of the outer segments of rod and cone cells. The outcome is a generator potential which, if it exceeds 'threshold strength', results in an action potential being propagated by the optic nerve to the brain

At the synapses between photoreceptor cells and neurones of the retina, action potentials are transmitted directly across the synaptic cleft (an **electrical synapse**). This type of synapse is not affected by drugs and is not susceptible to fatigue.

Images of our environment are reduced to two-dimensional images on the surface of the retina. Action potentials are thus sent via the optic nerve to the **visual cortex** of the brain. The brain receives action potentials from the left and right visual fields separately, and combines these to produce a single impression, our sight. This 'seeing process', known as **perception**, is complex. A good introduction to the investigation of perception is given in the *New Scientist*, Inside Science Number 99, 'Before your very eyes'; it is well worth reading.

6 In bright light the pupil diameter decreases; in dim light it increases. How does variation in pupil diameter also influence accommodation?

Organisation of the nervous system

The nervous system has two components with distinctive, but linked, roles:

- coordination and control of the activities of the body, carried out largely by the central nervous system (CNS), which is composed of the spinal cord and the brain;
- quick, precise communication between all parts, carried out by the peripheral nervous system (PNS).

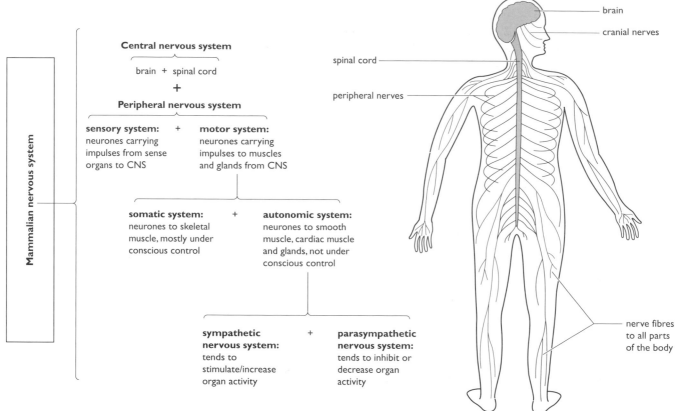

Figure 8.19 The organisation of the mammalian nervous system.

A **reflex action** is a rapid, automatic response involving a reflex arc (page 59); it may or may not involve conscious awareness. A reflex action begins with an action potential initiated by a stimulus, and involves sensory neurones, CNS and motor neurones.

Figure 8.20 Illustration of a reflex action.

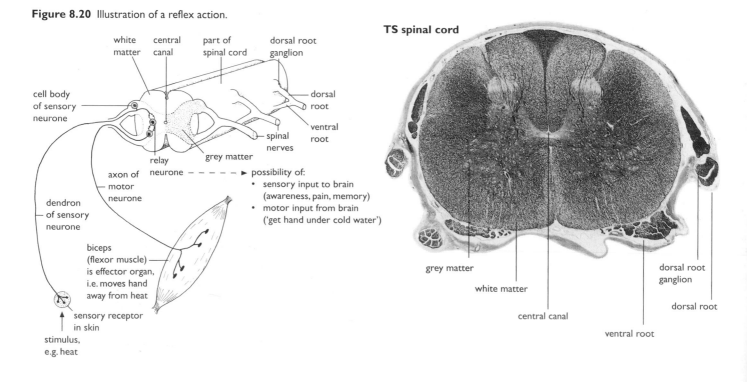

The **brain** controls all body functions, except those under the control of simple spinal reflexes. The human brain contains about 10^{11-12} relay neurones and a similar number of supporting cells (neuroglia cells). Each relay neurone is in synaptic contact with about 1000 other neurones.

In the embryo the brain develops from the neural tube. The anterior end of the tube enlarges to form three primary brain vesicles, the fore-, mid- and hindbrain. The various parts of the mature brain develop from these vesicles by selective thickening and folding processes in the walls and the roof. As in the spinal cord, white and grey matter are also present in the brain. Grey matter (cell bodies and synapses) makes up the interior of the brain; white matter (myelinated nerve fibres) occurs towards the exterior of the brain. In the development of the mammalian brain, grey matter forms an additional layer on the cerebral hemispheres and cerebellum. Cerebral blood vessels have walls that act as a barrier to many of the dissolved substances in the blood. However, oxygen and glucose cross this **blood/brain barrier** without restriction.

> **7** What structural differences would you expect in the reflex arc of a reflex action when the brain **a)** may be involved, and **b)** is not involved?

Parts of the brain:

1 **Hypothalamus:** this forms part of the floor of the forebrain, and is well supplied with blood vessels. It is the main control centre for the autonomic nervous system. Centres in the hypothalamus regulate body activities concerned with maintaining a constant internal environment (body temperature, the levels of metabolites such as sugars, amino acids and ions, and hormones in the blood), feeding and drinking reflexes, and aggressive and reproductive behaviour. The hypothalamus is the main link between the nervous and endocrine systems. It is attached to the pituitary gland and controls its release of hormones.

2 **Cerebral hemispheres:** these are extensions of the forebrain and form the bulk of the human brain. The hemispheres have a vastly extended surface, achieved by foldings with deep grooves. The surface is covered by grey matter up to a depth of 3 mm, known as the cerebral cortex. It is densely packed with non-myelinated neurones. The areas of the cortex with special sensory and motor functions have been mapped out; in these areas the body's voluntary and the majority of involuntary activities are coordinated.

3 **Cerebellum:** the two hemispheres of the cerebellum have a folded external surface layer of grey matter. The cerebellum is concerned with the control of involuntary muscle movements of posture and balance, and of precise, voluntary manipulations involved in hand work, speech and writing. These movements are coordinated, rather than initiated, here.

4 **Medulla:** the regulatory centres here are concerned with automatic adjustment of heart rate, ventilation of the lungs, temperature regulation and with swallowing, salivation, vomiting, coughing and sneezing. Sensory information arrives via the vagus nerve. In the medulla, the ascending and descending tracts of nerve fibres connecting the spinal column and brain cross over, so that the left side of the body is controlled by the right side of the brain, and *vice versa*.

Figure 8.21 The human brain.

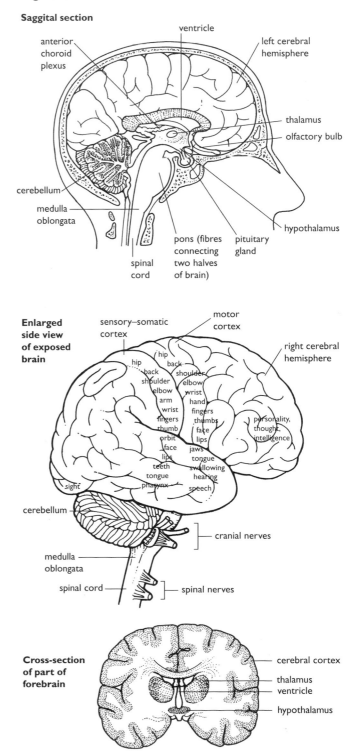

The endocrine system

The endocrine system plays a part in controlling the body by releasing chemical messengers called **hormones**. These are secreted by exocytosis directly into the bloodstream, and are transported indiscriminately all over the body. However, they only act at specific sites, called the target organs. Most hormones cause changes to the metabolic reactions of their target organs before being excreted from the body. Examples of hormones already discussed include cholecystokinin (pancreozymin, page 22), antidiuretic hormone (ADH, page 46) and insulin (page 50).

Table 8.1 Endocrine and nervous systems compared.

Endocrine system	Nervous system
Communication by chemical messengers transmitted in the bloodstream	Communication by electrochemical action potentials (impulses) transmitted via nerve fibres
Hormones 'broadcast' all over the body but influence target cells and tissues only	Action potentials are targeted on specific cells
Causes changes in metabolic activity	Causes muscles to contract or glands to secrete
Effects may occur over many minutes, several hours or longer	Effects occur within milliseconds
Effects tend to be long lasting	Effects tend to be short-lived and reversible

The pituitary, the master endocrine gland, develops in the embryo by amalgamation of an outgrowth of the roof of the mouth (forming the anterior pituitary) with an outgrowth of the floor of the forebrain below the hypothalamus (forming the posterior pituitary). The hypothalamus and pituitary gland work together in two ways:

1 the hypothalamus secretes hormones that regulate aspects of pituitary function;
2 the hypothalamus monitors the levels of hormones in the blood and exerts negative feedback control (i.e. when hormones are at a low level their secretion is stimulated; when high, their secretion is inhibited).

The **anterior pituitary** lobe produces and secretes hormones that regulate other endocrine glands:

- thyroid-stimulating hormone (TSH), which controls the secretion of hormones by the thyroid gland. The thyroid gland lies in the neck and produces hormones that regulate general body metabolism and the rate of glucose metabolism;
- adrenocorticotrophic hormone (ACTH), which regulates the secretion of hormones by the adrenal cortex (see opposite);
- follicle-stimulating hormone (FSH), which stimulates spermatogenesis (in the testes) and growth of follicles (in the ovaries) (page 82);
- luteinising hormone (LH), which affects hormone production by the testes and ovaries (pages 79 and 82);
- somatotrophin, a growth hormone that regulates the production of other hormones by the liver.

Figure 8.22 The pituitary gland in LS (x 50).

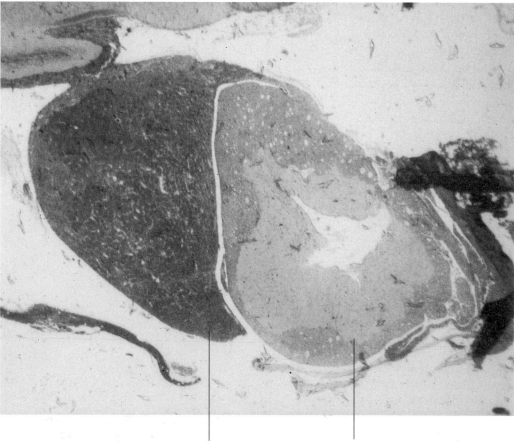

anterior lobe

posterior lobe

Secretions by the anterior lobe are controlled by two types of hormone released from the hypothalamus, known as 'releasing' and 'inhibiting' hormones.

The **posterior pituitary** lobe releases two hormones: ADH (page 46) and oxytocin. Both of these hormones are produced in the hypothalamus and stored in the posterior lobe of the pituitary until required.

The **adrenal glands** lie immediately above the kidneys. These glands consist of two quite distinct regions, an outer adrenal cortex (the bulk of the gland) and an inner adrenal medulla.

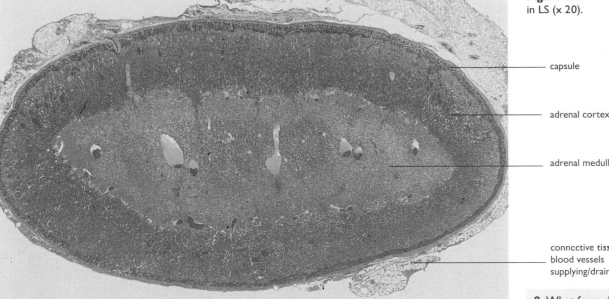

Figure 8.23 The adrenal gland in LS (x 20).

capsule

adrenal cortex

adrenal medulla

connective tissue with blood vessels supplying/draining gland

Secretions of the adrenal cortex are triggered by hormones from the anterior pituitary and of the medulla by the sympathetic nervous system.

The **adrenal cortex** uses blood cholesterol to produce hormones known as the corticoids. Corticoid hormones are relatively slow acting. They fall into two groups:

1 glucocorticoids, which are produced in periods of anxiety or disease, and cause glucose and glycogen to be formed from fats and proteins;
2 mineralocorticoids, which are involved in water and ion regulation, e.g. aldosterone (page 46).

8 What form does the 'flight or fight' response actually take?

9 What roles are fulfilled by blood capillaries supplying glands like the adrenals?

The **adrenal medulla** produces two hormones, adrenaline and noradrenaline, which augment the 'flight or fight' response. This prepares the body for exertion and high physical and mental performance.

The two modes of action of hormones

Chemical analysis of hormones has shown that they fall into two groups, the **steroid hormones** (e.g. oestrogen, testosterone, aldosterone) and the **amine/peptide hormones** (e.g. ADH, TSH, insulin). These two groups have different modes of action on target cells.

Figure 8.24 How hormones influence target cells.

Amine/peptide hormone action

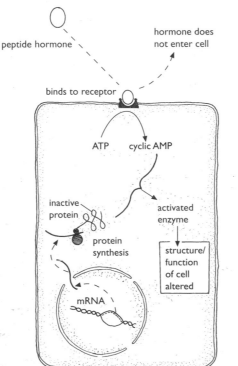

peptide hormone

hormone does not enter cell

binds to receptor

ATP   cyclic AMP

inactive protein

activated enzyme

protein synthesis

structure/ function of cell altered

mRNA

Steroid hormone action

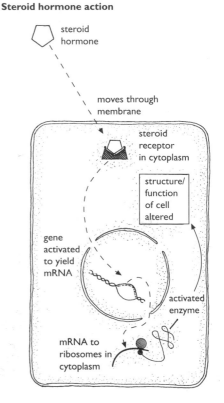

steroid hormone

moves through membrane

steroid receptor in cytoplasm

structure/ function of cell altered

gene activated to yield mRNA

activated enzyme

mRNA to ribosomes in cytoplasm

Support and locomotion

Living things are able to support themselves and maintain positions in which essential processes of life, such as nutrition, can be carried out. A mammal is supported and partially protected by its skeleton, which works in association with the musculature to give the body its permanent shape and form. In addition, the skeleton acts as a system of levers facilitating locomotion. Mammals have a jointed internal skeleton (an **endoskeleton**) of bone and some cartilage. Their skeleton contrasts with the external or **exoskeleton** of insects, and the hydrostatic skeleton of organisms such as the earthworm.

Bone consists of a hard, calcified matrix with numerous specialised bone cells (osteocytes). These cells occur in tiny cavities (lacunae) in the matrix, with their cytoplasmic connections running in tiny tubes. Chemically, bone is approximately 70% mineral matter, mainly calcium phosphate, and 30% organic matter, of mucopolysaccharide and collagen fibres (and cells). Bone cells secrete *and* maintain the matrix. The matrix is impervious to tissue fluid, but blood vessels permeate the bone tissue in channels, delivering essential nutrients and removing waste products.

There are two types of bone tissue. **Compact** bone is very dense, and makes up the greater part of the bones of the body. **Spongy** bone occurs within the larger bones, and contains large spaces occupied by bone marrow.

1 Which internal organs of mammals can be said to be fully or partly protected by the skeleton?

Figure 9.1 The structure of bone.

Mammalian femur in LS

Microscopic structure of compact bone

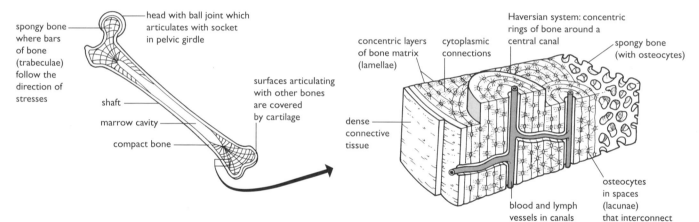

TS of compact bone, HP (x 200)

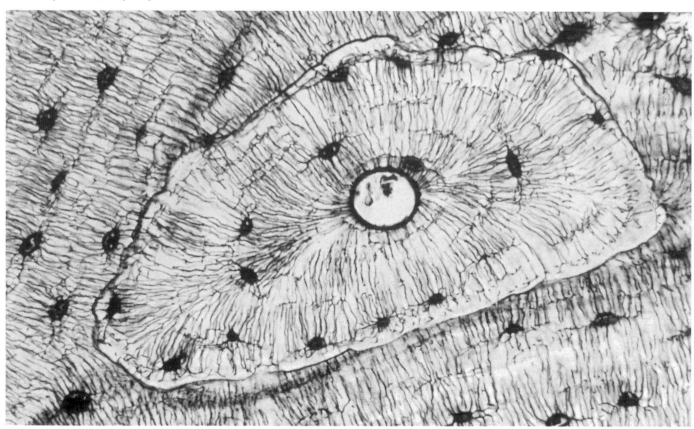

Cartilage is a firm, flexible tissue, elastic and resistant to strain. In the adult mammal, cartilage is restricted to the surface of bones at joints, the flexible connections between the ribs and sternum, the incomplete rings that hold open the trachea and bronchi, the outer ear, tip of the nose, and the epiglottis. Special fibre-reinforced cartilage makes up the intervertebral discs of the spinal cord.

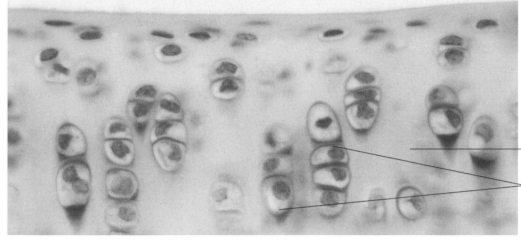

In cartilage the matrix is made of a protein, **chondrin**. This matrix is secreted by cells, chondrocytes, that become embedded in the matrix in small clusters. Blood vessels do not penetrate; nutrients are supplied and waste products are lost by diffusion. This type of cartilage is called **'hyaline'**.

— matrix

— groups of chondrocytes

Figure 9.2 TS cartilage, HP (x100).

Movement at joints

Muscles for locomotion are attached to the movable parts of skeletons across joints. Attachment is typically by tendons, which are cords of fibrous connective tissue. For movement at the joint to occur, one end of the muscle must be anchored to bone that in this context is non-movable. Muscle fibres are able to shorten by a half to a third of their 'resting' length, but they cannot expand. However, skeletal muscles occur in antagonistic pairs (Figure 9.11, page 77), so that when one member of a pair contracts, the other is stretched. Joints in the skeleton occur where bones meet. Immovable joints can be seen in the bones of the cranium, but most joints permit controlled movement of bones. In the leg, there is a ball-and-socket joint at the hip, and a hinge joint at the knee.

2 Apart from voluntary (skeletal) muscle, what other types of muscle occur in the body?

Figure 9.3 The functioning knee joint.

Flexed knee joint of guinea pig in LS, LP

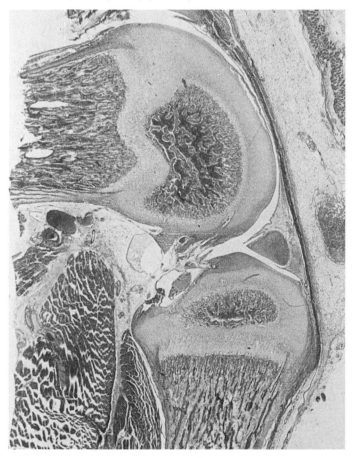

Knee joint drawn in saggital section

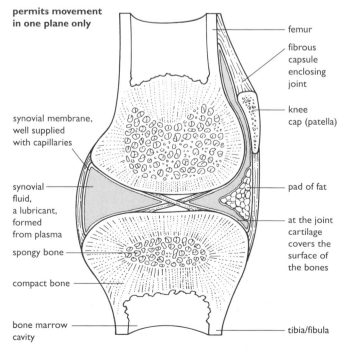

permits movement in one plane only

synovial membrane, well supplied with capillaries

synovial fluid, a lubricant, formed from plasma

spongy bone

compact bone

bone marrow cavity

femur

fibrous capsule enclosing joint

knee cap (patella)

pad of fat

at the joint cartilage covers the surface of the bones

tibia/fibula

Voluntary muscle: structure and function

Voluntary (skeletal) muscle mainly occurs attached to the skeleton in mammals. Voluntary muscle consists of many thousands of elongated, cylindrical, multinucleated muscle fibres in parallel bundles, bound by connective tissue. Each fibre has many nuclei. Blood vessels run beside the fibres and serve numerous capillary networks. Under the light microscope, the muscle fibres are seen to have alternate light and dark bands or striations, so voluntary muscle is also referred to as **striped or striated** muscle.

3 What are multinucleate structures such as this called?

Figure 9.4 The structure of voluntary muscle.

Voluntary muscle cut to show the bundle of fibres

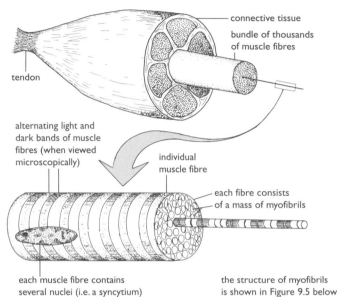

connective tissue

bundle of thousands of muscle fibres

tendon

alternating light and dark bands of muscle fibres (when viewed microscopically)

individual muscle fibre

each fibre consists of a mass of myofibrils

each muscle fibre contains several nuclei (i.e. a syncytium)

the structure of myofibrils is shown in Figure 9.5 below

LS voluntary muscle, HP (phase contrast) (x 1500)

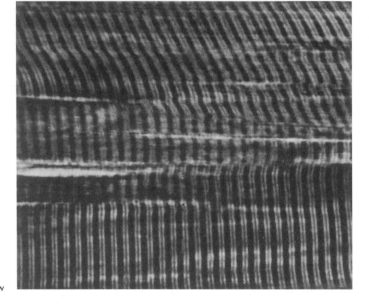

   The ultrastructure of striped muscle can be shown by an electronmicroscope. A muscle fibre consists of a mass of tiny, parallel **myofibrils**, running the length of the fibre. These are contained within the plasma membrane (the sarcolemma) of the fibre, surrounded by cytoplasm (sarcoplasm) containing endoplasmic reticulum and numerous mitochondria. The complex pattern of light and dark bands and stripes is due to the presence of a system of interlocking protein filaments of two types:

- **thick filaments**: these are shorter than the thin filaments, and are composed of many myosin molecules. Each myosin molecule has a bulbous head, forming a cross-bridge;
- **thin filaments**: these extend between the thick filaments and are held together by transverse bands (Z bands). Thin filaments are composed of actin, together with an additional protein.

For convenience of description each repeating unit of the myofibril is referred to as a **sarcomere**.

Figure 9.5 TEM of striped muscle (x 4100) with interpretive diagram.

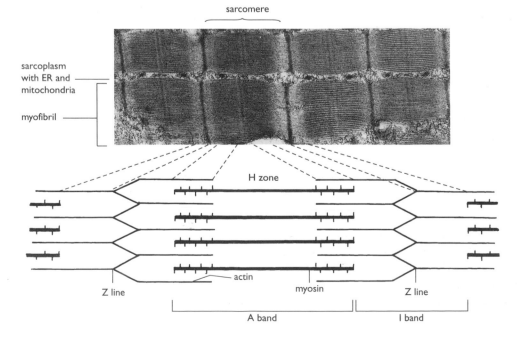

sarcomere

sarcoplasm with ER and mitochondria

myofibril

H zone

Z line

actin

myosin

Z line

A band

I band

Muscle fibres are **innervated** by a motor neurone nerve ending, at a structure known as a motor end plate. A motor end plate is a special type of synapse (page 62).

Figure 9.6 Motor end plates.

The **sliding-filament hypothesis** suggests that muscle contraction occurs when the thin protein filaments (actin) are slid inwards. This is brought about by the action of myosin cross-bridges becoming temporarily attached to the adjacent actin; the complex swings through an angle and pulls the actin towards the centre of the sarcomere. This movement is likened to a 'ratchet mechanism'.

Figure 9.7 The sliding-filament hypothesis.

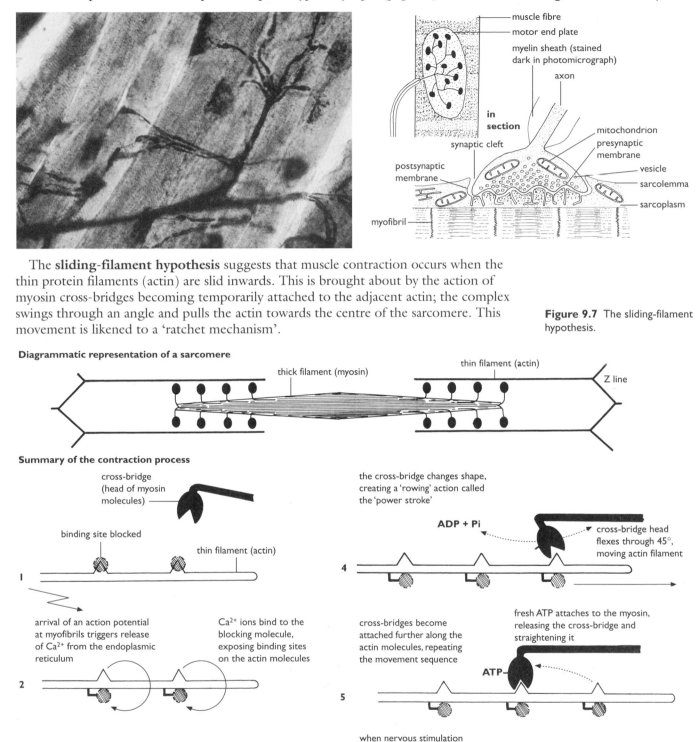

Diagrammatic representation of a sarcomere

thick filament (myosin)

thin filament (actin)

Z line

Summary of the contraction process

cross-bridge (head of myosin molecules)

the cross-bridge changes shape, creating a 'rowing' action called the 'power stroke'

binding site blocked

thin filament (actin)

ADP + Pi

cross-bridge head flexes through 45°, moving actin filament

1

4

arrival of an action potential at myofibrils triggers release of Ca²⁺ from the endoplasmic reticulum

Ca²⁺ ions bind to the blocking molecule, exposing binding sites on the actin molecules

cross-bridges become attached further along the actin molecules, repeating the movement sequence

fresh ATP attaches to the myosin, releasing the cross-bridge and straightening it

ATP

2

5

head of the cross-bridge attaches

ATP

when nervous stimulation ceases, Ca²⁺ ions return to the endoplasmic reticulum and the binding sites are blocked again

contracted sarcomeres are 're-set' for contraction by being stretched, by the action of an antagonistic muscle

3

6

7

Resources for contraction come as fatty acids, glucose and oxygen from the blood. **Glycogen** is held in reserve in muscle fibres and is available for hydrolysis to glucose when required. Aerobic respiration of glucose produces ATP. In repeated contraction, oxygen and glucose are used up faster than they can be delivered. Then muscle makes use of oxygen reserves, held as **oxymyoglobin**, and reserve ATP, generated from **creatine phosphate**. If muscle continues to be active, such as in prolonged heavy work, anaerobic respiration by lactic acid fermentation becomes the only source of ATP.

Muscles as effectors

I Posture and balance

The relative position and state of tension of the whole body musculature is kept under continuous, unconscious review, by the brain. Body position is maintained with the help of sensory data, particularly that received from three types of internal receptors.

- **Proprioceptors** are tiny sense organs and sensitive nerve endings located in the capsules of joints, on the surface of tendons and in other body tissues. Local movements and even slight changes in tension stimulate these receptors to send impulses to the brain via the spinal cord. Changes in localised pressure are detected by pressure receptors (Figure 8.12, page 64) situated in various body tissues.

- **Muscle spindles** are stretch receptors found in skeletal muscles. If the body tends to sag under its own weight, for example, nerve endings in the spindle are further stretched and fire additional impulses to the spinal cord, triggering the stretch reflex to make the necessary adjustment to muscle tone and hence body posture. Impulses are also passed on to the brain, via the spinal cord.

- The **ear** performs two sensory functions, hearing and 'balance'. The ear drum is a thin, strong sheet of elastic connective tissue that vibrates in response to sound waves. The ear bones (ossicles) of the middle ear act as levers, transmitting vibrations to the oval window and magnifying the vibrations twenty-fold as they do so. The inner ear is a fluid-filled cavity. The lower compartment contains the **cochlea**, which transduces mechanical vibrations to nerve impulses that are relayed to the brain and used to give us the sensations of sound. Above the cochlea, but connected to it, there lies a system of fluid-filled canals and sacs, known as the **vestibular apparatus**.

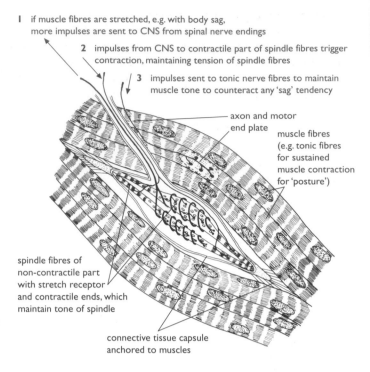

1 if muscle fibres are stretched, e.g. with body sag, more impulses are sent to CNS from spinal nerve endings

2 impulses from CNS to contractile part of spindle fibres trigger contraction, maintaining tension of spindle fibres

3 impulses sent to tonic nerve fibres to maintain muscle tone to counteract any 'sag' tendency

axon and motor end plate

muscle fibres (e.g. tonic fibres for sustained muscle contraction for 'posture')

spindle fibres of non-contractile part with stretch receptor and contractile ends, which maintain tone of spindle

connective tissue capsule anchored to muscles

Figure 9.8 Muscle spindles are found in skeletal muscles.

Figure 9.9 Structure of the ear.

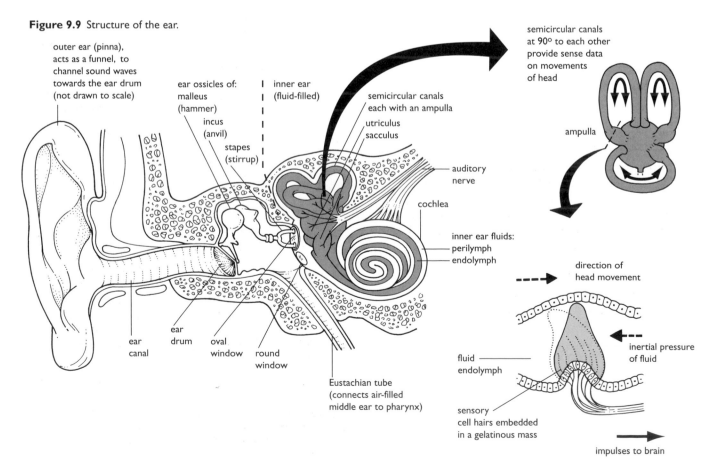

outer ear (pinna), acts as a funnel, to channel sound waves towards the ear drum (not drawn to scale)

ear ossicles of:
malleus (hammer)
incus (anvil)
stapes (stirrup)

inner ear (fluid-filled)

semicircular canals each with an ampulla
utriculus
sacculus

auditory nerve

cochlea

inner ear fluids:
perilymph
endolymph

ear canal

ear drum

oval window

round window

Eustachian tube (connects air-filled middle ear to pharynx)

semicircular canals at 90° to each other provide sense data on movements of head

ampulla

direction of head movement

inertial pressure of fluid

fluid endolymph

sensory cell hairs embedded in a gelatinous mass

impulses to brain

The **vestibular apparatus** of the ear consists of the three semicircular canals, each with an ampulla (to detect movement of the head), and the utriculus and sacculus (to detect posture). The receptor cells in the vestibular apparatus (found in the ampullae and in the walls of the utriculus and sacculus) have hair-like extensions embedded in dense structures that are supported in a fluid called endolymph. Movements may cause deflection of the hairs and the generation of an action potential.

Figure 9.10 Macula (posture detection) from the floor of the sacculus, HP.

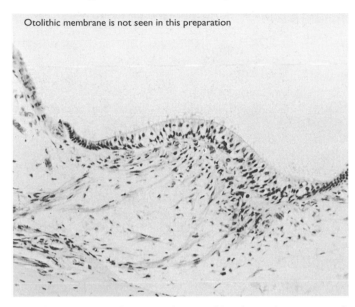

Otolithic membrane is not seen in this preparation

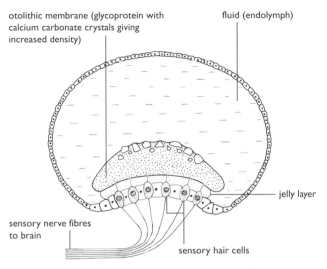

otolithic membrane (glycoprotein with calcium carbonate crystals giving increased density)

fluid (endolymph)

jelly layer

sensory nerve fibres to brain

sensory hair cells

movements of the otolithic membrane pull on the sensitive hairs, stimulating the sensory cells, and causing action potentials to be sent to the brain

Other sensory data are also used in the maintenance of balance. Observe what happens when you try balancing on one leg for a few seconds, first with your eyes open, then with them closed! All incoming sensory data, help update the brain on body position and movement in relation to internal and external conditions.

There are **two types of voluntary muscle fibre** with different physiological properties, known as **tonic** (slow) fibres and **twitch** (fast) fibres. Twitch fibres are fast contractors (they reach maximum tension following a single stimulus), and are used in locomotion. Tonic fibres allow the sustained muscular contraction needed to maintain body posture. Whilst we are awake, tonic fibres of our voluntary muscles are in a permanent state of sustained tension.

2 Walking as a series of reflex actions

An animal moves forwards by pushing backwards against its surroundings. Terrestrial mammals make a push downwards and slightly backwards on the ground. Pairs of antagonistic muscles are responsible for propulsion, many pairs being involved in the movement of each limb. When we walk we use about 60 muscles in each leg and the related part of the hips. Figure 9.11 is a simplified drawing of the main muscles involved in leg movements. Can you feel each of these in your leg as it contracts as part of normal walking movements? The labels in the drawing give clues about the role of each muscle.

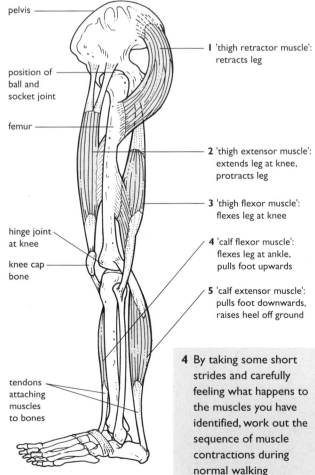

pelvis

position of ball and socket joint

femur

hinge joint at knee

knee cap bone

tendons attaching muscles to bones

I 'thigh retractor muscle': retracts leg

2 'thigh extensor muscle': extends leg at knee, protracts leg

3 'thigh flexor muscle': flexes leg at knee

4 'calf flexor muscle': flexes leg at ankle, pulls foot upwards

5 'calf extensor muscle': pulls foot downwards, raises heel off ground

4 By taking some short strides and carefully feeling what happens to the muscles you have identified, work out the sequence of muscle contractions during normal walking movements. Write these down in sequence, explaining the consequence of each contraction in walking.

Figure 9.11 Lateral view of the left human leg, showing the position of key muscles used in walking.

10 Reproduction

Sexual reproduction involves the production of special sex cells (**gametes**) and their fusion to form a **zygote**, which then grows and develops to form a new individual. Mammals are unisexual. Gametes (egg cells or sperm) are produced in paired organs called **gonads** (ovaries or testes), the **primary sexual organs**. A system of tubes and glands help in the storage and transfer of gametes, and these make up the **secondary sexual organs**.

Male reproductive system

The **male reproductive system** produces sperm in the testes and delivers them in a liquid medium, called semen, to the vagina during copulation. The testes also produce the chief male sex hormone, **testosterone**, in the interstitial cells.

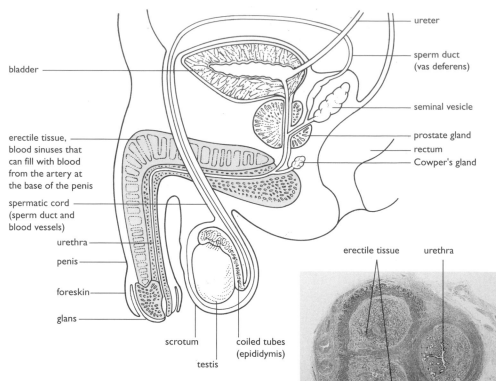

Figure 10.1 Reproductive organs of the human male. The inset shows TS of the penis, LP.

In most species of adult mammal, the **testes** are suspended in an external bag called the scrotum, even though they develop in the abdomen. The human testes are oval-shaped, about 3 cm in length. Each consists of about a thousand seminiferous tubules held together by connective tissue that contains the interstitial cells and blood capillaries. Sperm production occurs in the **seminiferous tubules**. Each tubule is connected to the epididymis, which leads into a sperm duct. The sperm ducts fuse with the urethra just below the bladder. Here the prostate gland occurs, and nearby are found the seminal vesicles and Cowper's gland. These glands contribute the fluid part of semen.

1 In what ways does asexual reproduction fundamentally differ from sexual reproduction?

TS of seminiferous tubules, LP

Dissected to show components

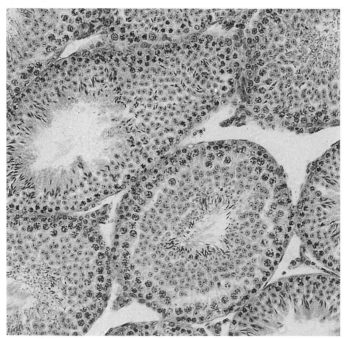

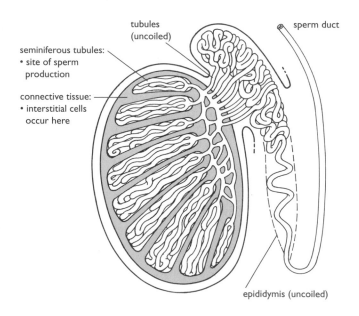

Figure 10.2 The structure of the testis.

Sperm production (spermatogenesis) starts at puberty. The seminiferous tubules are lined by **germinal epithelial cells**, which remain attached to the basement membrane but divide repeatedly by mitosis, releasing cells called **spermatogonia**. These divide mitotically and grow in size, forming **primary spermatocytes**. Next, by the reductive division of meiosis, haploid **spermatids** are formed. Elongated nutritive cells are found between the germinal epithelium cells. The spermatids position themselves here and develop into sperm.

2 What condition, experienced outside the abdominal cavity, is necessary for the proper functioning of the testes?

Part of the seminiferous tubule, seen in TS

Steps to sperm formation

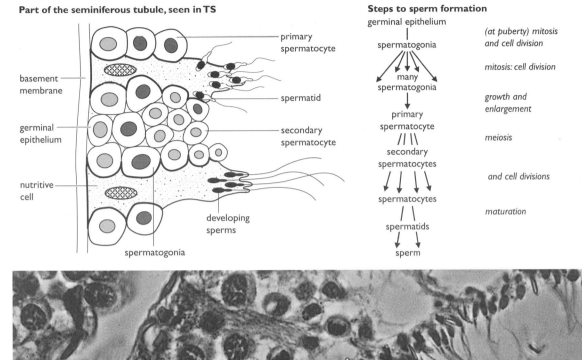

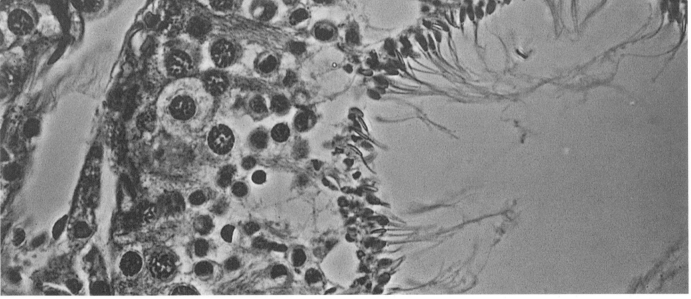

Figure 10.3 Sperm production (spermatogenesis).

The **sperm** progressively develop the power of mobility as they are moved from the seminiferous tubules to the epididymis. They are retained in the epididymis for up to 10 days, continuing with their maturation processes. Finally, sperm are held in the sperm ducts prior to ejaculation (page 83).

Semen is a viscous, slightly alkaline fluid, rich in mucus. Enzymes that activate the sperm are also present.

3 List the male secondary sexual characteristics.

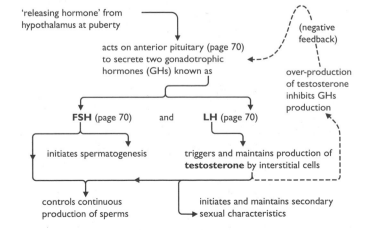

Figure 10.4 The testis as an endocrine gland – a summary.

Female reproductive system

The **female reproductive system** produces **egg cells** in the ovaries, and delivers them to the oviduct where fertilisation may occur. The wall of the uterus is prepared for implantation and development of an embryo. The ovaries also produce the chief female sex hormone, **oestrogen**.

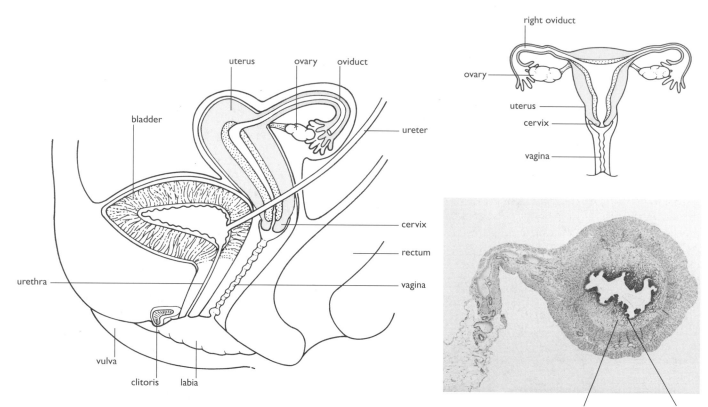

Figure 10.5 Reproductive organs of the human female. The inset shows TS of the uterus, LP.

The human **ovaries** are oval-shaped structures, about 3 cm long, held in the lower abdominal cavity by outfolds of the peritoneum (the lining of the abdomen). Beside each ovary is the funnel-shaped opening of the oviduct. The oviducts are thin-walled, muscular tubes with an interior lined by ciliated epithelium containing mucus-secreting cells. The oviducts lead into the top of the uterus.

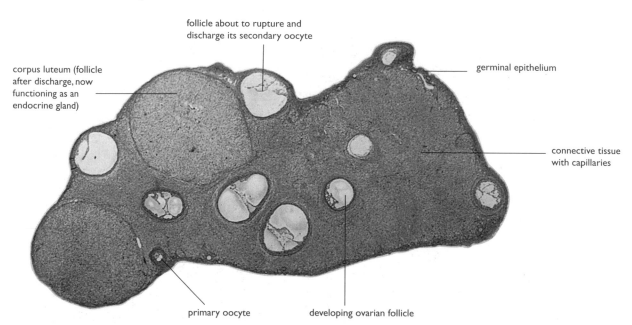

Figure 10.6 TS mammalian ovary (x 15).

Production of egg cells (oogenesis) occurs in the ovaries of the human fetus before birth. Cells of the **germinal epithelium** divide repeatedly by mitosis, forming numerous **oogonia**. In the ovary, oogonia develop into **primary oocytes**, which then become surrounded by follicle cells to form **primary follicles**. At birth, the human ovary typically contains about 200 000 tiny primary follicles. These remain dormant until puberty, and less than 1% ever complete further development.

Between puberty and menopause several primary follicles start to develop each month, but typically only one matures, ending in a position beside the outer wall of the ovary. During maturation of this primary oocyte, the first meiotic division occurs, but the cytoplasmic division that follows is unequal. A **secondary oocyte** and a tiny polar body are formed. Then, the secondary oocyte *commences* the second meiotic division; this division is only completed if fertilisation occurs. The secondary oocyte is released (**ovulation**) by rupture of follicle and ovary wall. The remains of the primary follicle develop into the **corpus luteum**, an additional but temporary endocrine gland. The release of one 'egg cell' approximately every 28 days (ovulation) is one event in the sequence known as the **menstrual cycle**. During this cycle, the lining of the uterus (endometrium) is built up to accept an embryo, providing a favourable environment for the development of a fetus. When **implantation** of an embryo does not occur, the expanded endometrium is broken down and shed (menstruation or a 'period'). The ovarian cycle then begins again.

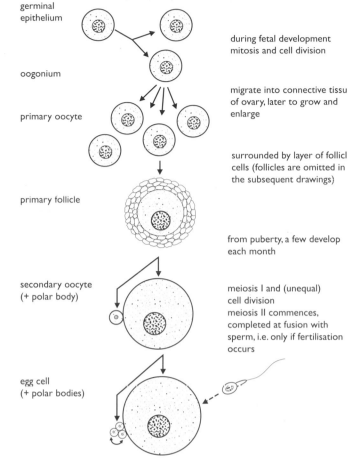

Figure 10.7 Steps to secondary oocyte formation.

germinal epithelium

during fetal development mitosis and cell division

oogonium

migrate into connective tissue of ovary, later to grow and enlarge

primary oocyte

surrounded by layer of follicle cells (follicles are omitted in the subsequent drawings)

primary follicle

from puberty, a few develop each month

secondary oocyte (+ polar body)

meiosis I and (unequal) cell division
meiosis II commences, completed at fusion with sperm, i.e. only if fertilisation occurs

egg cell (+ polar bodies)

4 Tabulate the significant differences between spermatogenesis and oogenesis.

Figure 10.8 The menstrual cycle.

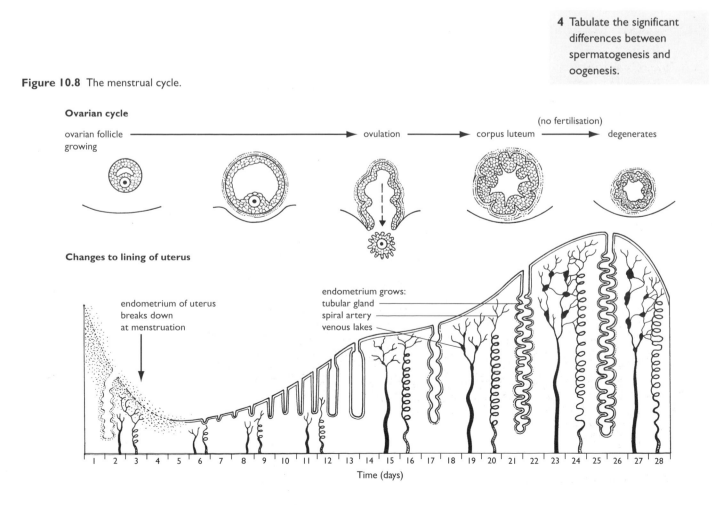

Ovarian cycle

ovarian follicle growing → ovulation → corpus luteum → (no fertilisation) degenerates

Changes to lining of uterus

endometrium of uterus breaks down at menstruation

endometrium grows:
tubular gland
spiral artery
venous lakes

Time (days)

Endocrine control of the menstrual cycle

The constantly changing levels of at least four hormones control the menstrual cycle. The anterior lobe of the pituitary secretes **FSH** and **LH**, both **gonadotrophic hormones**. The wall of the developing follicle secretes **oestrogen**. The vacant follicle (from which the secondary oocyte has been discharged) becomes the corpus luteum and secretes the hormone **progesterone**. *Note:* if fertilisation occurs and (later) implantation of an embryo, the embryo and developing placenta take over the maintenance of the endometrium (page 86).

5 List the structures that function as endocrine glands in the regulation of the menstrual cycle, noting the main roles of each.

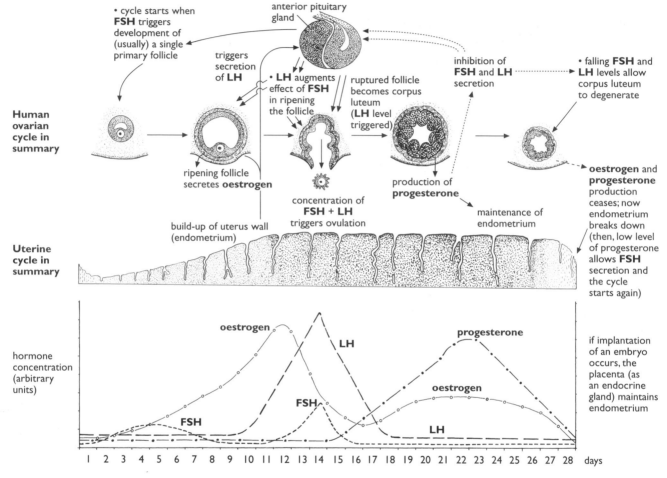

Figure 10.9 The changing levels of hormones in the regulation of the human menstrual cycle.

The **gametes of mammals** are oogamous, in that the male gamete is small and extremely motile, whilst the female gamete is large and sedentary. In humans, the egg cell and the head of the sperm are 100 μm and approximately 2.5 μm in diameter, respectively. Typically, female gametes contain a store of food that is used in development after fertilisation, but in mammals, food required for development is supplied through the placenta (page 85).

Unfertilised ovum (secondary oocyte) of rat (×150)

Spermatozoa of rat (×1500)

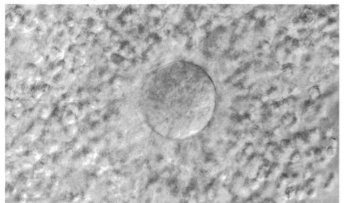

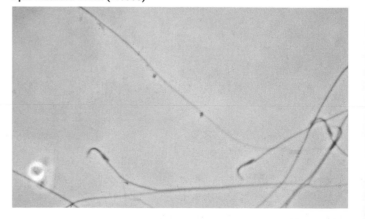

Figure 10.10 Photomicrographs of mammalian gametes.

In mammals, **fertilisation** is internal and occurs in the upper part of the oviduct. Semen is ejaculated from the erect penis close to the cervix. A volume of 3–5 cm^3 is released in humans. Waves of contractions in the muscular walls of the uterus and oviducts draw semen from the cervix to the site of fertilisation. One or more of the few sperm that reach the secondary oocyte have to pass between the follicle cells that surround the oocyte. Then, one sperm must traverse the glycoprotein coat (zona pellucida) to reach the egg for fertilisation to occur. Hydrolytic enzymes, packaged in the tip of the head of the sperm, digest a pathway.

Figure 10.11 Fertilisation of a human secondary oocyte.

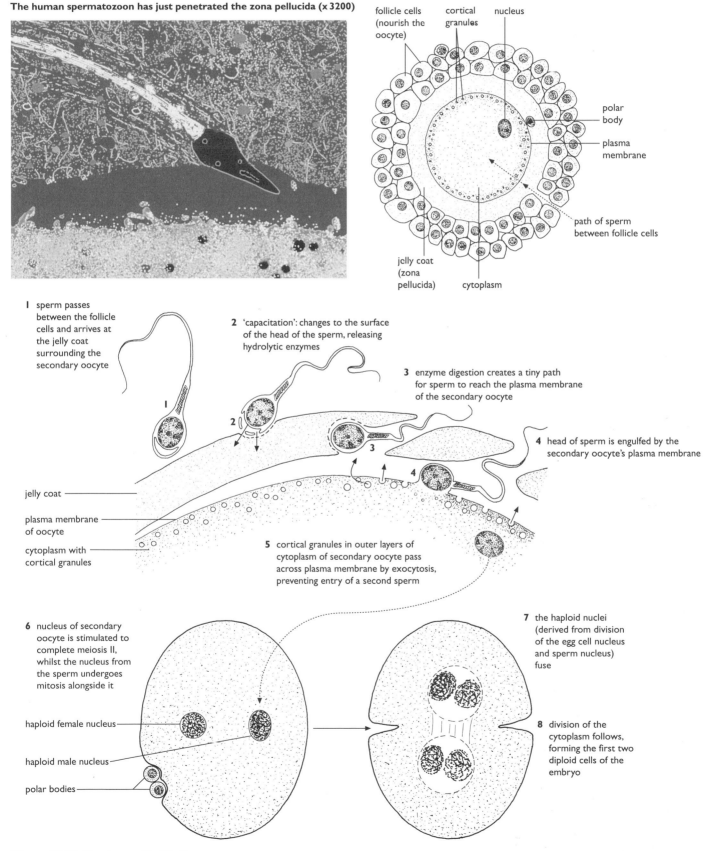

The human spermatozoon has just penetrated the zona pellucida (x 3200)

follicle cells (nourish the oocyte)

cortical granules

nucleus

polar body

plasma membrane

path of sperm between follicle cells

jelly coat (zona pellucida)

cytoplasm

1 sperm passes between the follicle cells and arrives at the jelly coat surrounding the secondary oocyte

2 'capacitation': changes to the surface of the head of the sperm, releasing hydrolytic enzymes

3 enzyme digestion creates a tiny path for sperm to reach the plasma membrane of the secondary oocyte

4 head of sperm is engulfed by the secondary oocyte's plasma membrane

jelly coat

plasma membrane of oocyte

cytoplasm with cortical granules

5 cortical granules in outer layers of cytoplasm of secondary oocyte pass across plasma membrane by exocytosis, preventing entry of a second sperm

6 nucleus of secondary oocyte is stimulated to complete meiosis II, whilst the nucleus from the sperm undergoes mitosis alongside it

7 the haploid nuclei (derived from division of the egg cell nucleus and sperm nucleus) fuse

haploid female nucleus

haploid male nucleus

polar bodies

8 division of the cytoplasm follows, forming the first two diploid cells of the embryo

Figure 10.12 The stages of fertilisation.

Gestation: zygote to embryo to fetus

Development in the uterus is known as **gestation** (40 weeks in humans). The rate of growth sustained in gestation far exceeds that in any other stage of life. In its first 2 months the offspring is described as an **embryo**, subsequently as a **fetus**. By the end of 2 months, the beginnings of the principle adult organs can be detected, and the embryonic membranes and placenta are operational.

Early embryo development starts as the zygote is transported down the oviduct by ciliary action. Division of the zygote into a mass of daughter cells is known as **cleavage**. The embryo does not increase in mass at this stage; by the time it reaches the uterus the embryo is a solid ball of tiny cells called **blastomeres**. Division continues, and the blastomeres organise themselves into a fluid-filled ball, the **blastocyst**.

Figure 10.14 The venue for the stage of cleavage in mammals.

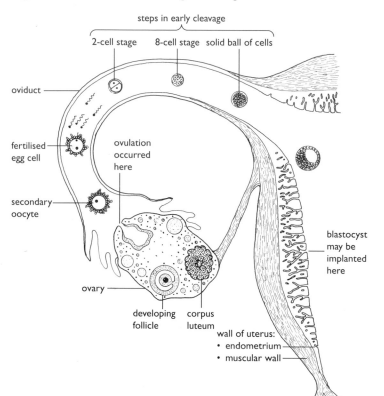

Figure 10.13 Cleavage in the rat (*Rattus*) embryo.

During cleavage, the number of cells increases but the size of the embryo does not increase

Two-cell stage

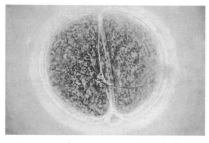

Four-cell stage

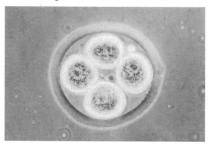

Eight-cell stage

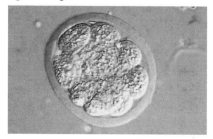

Early blastocyst

By day 7 in humans, the blastocyst (which is now about a hundred cells), starts to embed in the endometrium. This process, **implantation**, takes from day 7 to day 14. Some of the blastomeres become grouped as an **inner cell mass** that eventually forms the fetus proper. Once implanted, the embryo receives nutrients from the endometrium of the uterus

Figure 10.15 The embryo at implantation.

Blastocyst at about day 7 prior to implantation

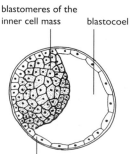

blastomeres of the inner cell mass

blastocoel

blastomeres of the trophoblast (first membrane system formed)

Implanted in the endometrium, about 14 days after fertilisation

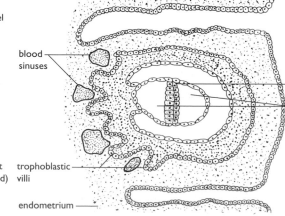

blood sinuses

trophoblastic villi

endometrium

cells that form the fetus proper

developing membrane systems (yolk sac plus amniotic cavity)

Cleavage is followed by **gastrulation** in which the inner cell mass rearranges to form distinct layers. Cell division continues, but is over-shadowed by coordinated movements of cells that form a complex embryo.

Subsequent developments are known as **organogeny**. Cells continue to divide whilst the **three germ layers** of the embryo appear. The innermost layer gives rise to the **endoderm** of the embryo, the outermost layer to the **ectoderm**, and in between is a cell mass that gives rise to the **mesoderm**: the embryo is said to be 'triploblastic'. The cells of the three germ layers develop into the tissues and organs of the embryo.

The placenta

The outer layers of the embryo give rise to the **membranes** and the placenta. One membrane, the amnion, surrounds a fluid-filled cavity in which the embryo is suspended, cushioned against mechanical damage. A second membrane forms the yolk sac (which has no obvious function in humans, but in reptiles and birds it is the nutrient store). Another membrane, the allantois, combines with part of the amnion and with the yolk sac to form the placenta. By the twelfth week the fetus is exchanging nutrients and waste products with the maternal blood system through a fully formed placenta.

The **placenta** is a disc-shaped structure of maternal (endometrial) and fetal membrane tissues. Here the maternal and fetal blood circulations are brought very close together but they do not mix. Exchange is by diffusion and active transport. The placenta starts to develop when the trophoblastic villi (see Figure 10.15) are invaded by blood vessels of the developing fetal circulation; it is built up as the embryonic membranes form. The placenta is also an endocrine gland, initially producing **human chorionic gonadotrophin (HCG)** (page 86) and later producing progesterone and oestrogen.

Figure 10.16 The origin and role of the three germ layers of the gastrula.

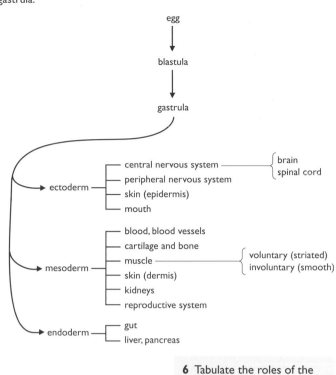

6 Tabulate the roles of the placenta, and describe briefly how these occur/are carried out.

Figure 10.17 The structure of the placenta.

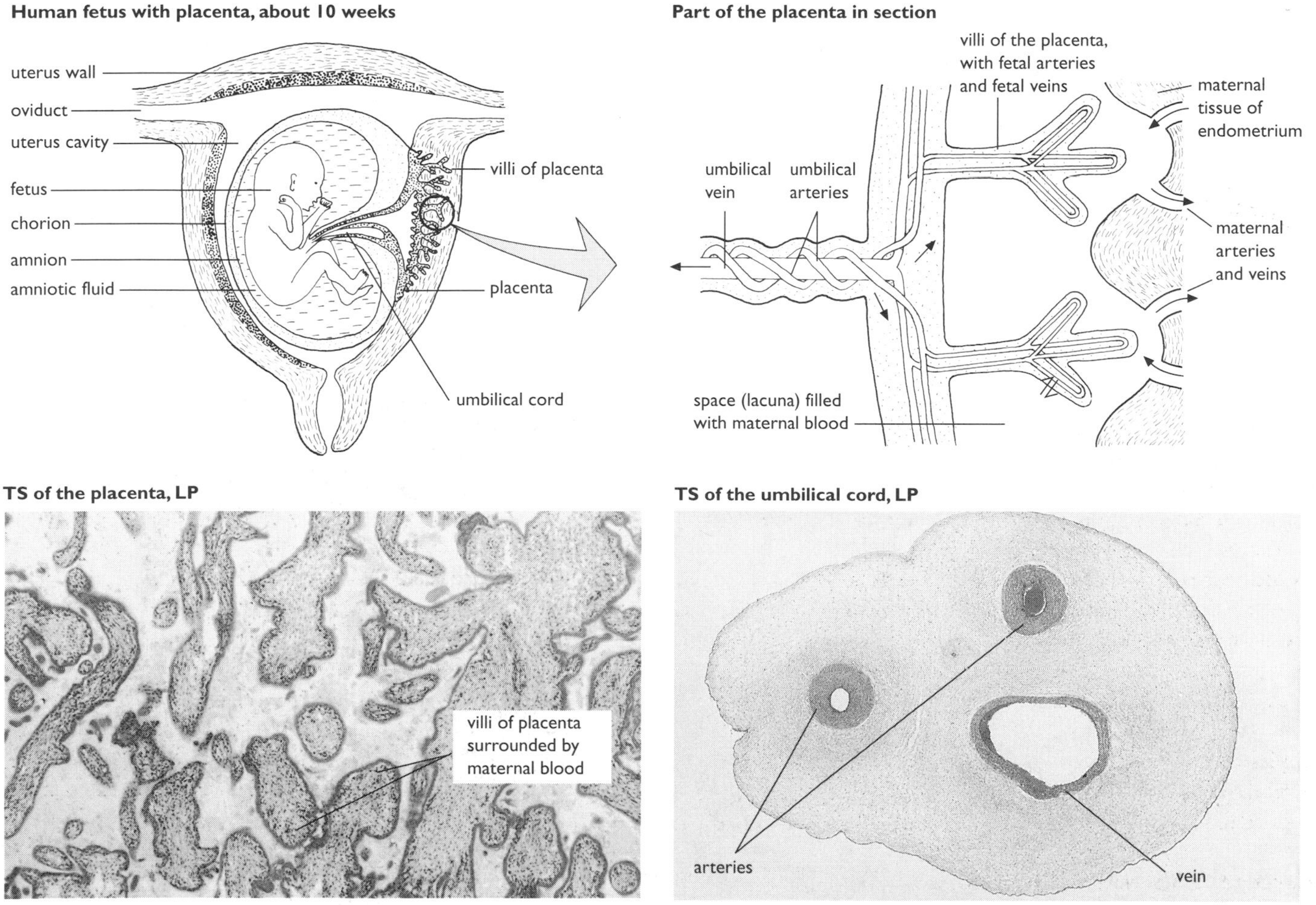

Human fetus with placenta, about 10 weeks

uterus wall
oviduct
uterus cavity
fetus
chorion
amnion
amniotic fluid
villi of placenta
placenta
umbilical cord

Part of the placenta in section

villi of the placenta, with fetal arteries and fetal veins
maternal tissue of endometrium
umbilical vein
umbilical arteries
maternal arteries and veins
space (lacuna) filled with maternal blood

TS of the placenta, LP

villi of placenta surrounded by maternal blood

TS of the umbilical cord, LP

arteries
vein

In the **hormonal control of pregnancy** an additional sex hormone, **HCG**, is produced. It appears in the urine about 7 days after conception. It is initially secreted by the blastomeres, but is later secreted by the placenta. HCG maintains the corpus luteum as an endocrine gland (secreting oestrogen and progesterone) for the first 16 weeks of pregnancy. When the corpus luteum eventually does break down, the placenta itself secretes oestrogen and progesterone. Without maintenance of these hormone levels a spontaneous abortion would occur.

In preparation for **birth** a third hormone, **relaxin**, is released by the placenta. This hormone relaxes the elastic fibres that join the bones of the pelvic girdle, thus aiding dilation of the cervix for the head (the widest part of the offspring) to pass through. At this time, powerful, intermittent waves of contraction of the muscles of the uterus wall start at the top of the uterus and end at the cervix. During 'labour', the rate and strength of the contractions increase to expel the offspring. Finally, lesser uterine contractions separate the placenta from the endometrium, and cause the discharge of the placenta and remains of the umbilicus as 'afterbirth'.

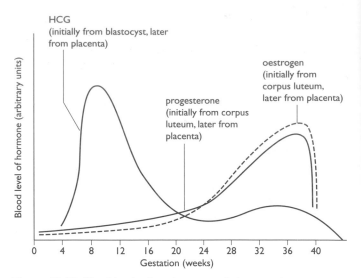

Figure 10.18 Blood levels of sex hormones during gestation.

Lactation is the production, secretion and ejection of **milk**. It is controlled by a hormone, prolactin, produced in the anterior pituitary gland. The mammary glands are first prepared for milk production by the action of progesterone and oestrogen. When the level of progesterone in the body falls abruptly just before birth, lactation commences. In the first 2–3 days the milk is called colostrum and is of special composition: sugar and protein, no fat, but with antibodies that aid survival during first exposures to potentially dangerous microorganisms. Milk itself is almost a complete diet (1.5–2.0% protein; 3.5% fat; 6.5% lactose (sugar); 0.3% minerals e.g. Ca^{2+}; vitamins A, B, C and D, and water. Milk is deficient in iron, and offspring have to rely on iron stored in the liver until the diet changes).

Figure 10.19 The mammary glands.

TS lactating mammary gland (mouse), HP

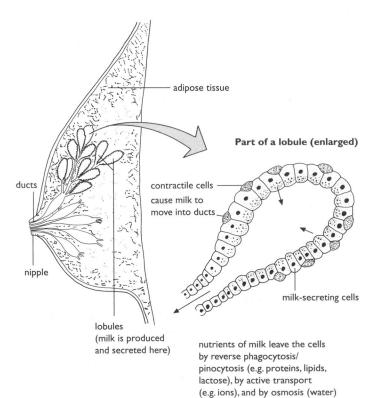

Answers

1 Introduction to classification and diversity

1 Phylum: post-anal tail; Class: skin covered by hair (or presence of mammary glands); Order: powerful jaws, with large canine teeth.

2 Cells and tissues

1 Cell theory concepts:
(i) building blocks of structure of living things; (ii) derived from other cells by division; (iii) contain information ('blue-print') for growth, development and functioning; (iv) site of the chemical reactions of life (metabolism).

2 An enzyme is a biological catalyst; a substance that alters the rate (speeds it up) of a reaction without itself being 'used up' in the reaction. Enzymes are mostly proteins, but some are made up of RNA.

   Enzymes work by forming an enzyme–substrate complex at an active surface of the enzyme (very briefly), thereby lowering the 'activation energy' needed for the reaction to occur.

3

Bacterial cell (prokaryotic)	Mammalian cell (eukaryotic)
Extremely small, e.g. 0.5–10 μm (about the size of some organelles in eukaryotic cells)	Larger, e.g. 10–100 μm
Nucleus absent, circular DNA helix in cytoplasm	Nucleus with nuclear membrane and chromatin/chromosomes
Cell wall present, of mucopeptides	No cell wall present
Few organelles present	Many organelles present
Small ribosomes	Larger ribosomes

4 Protein: contains the elements C, H, O and N, and sometimes S. Macromolecules, built from a linear sequence of amino acids, condensed together by peptide linkages (primary structure), and arranged into a secondary structure (sheets or helices), and into a tertiary structure of a complex shape made fairly secure by internal bonds.

   Lipid: contains the elements C, H and O (as do carbohydrates) but the proportion of O is much less than in carbohydrates. Lipids are insoluble in water. They are a complex group, but simple lipids (fats and oils) are esters formed by the condensation reaction between fatty acids and glycerol (an alcohol) to form triglycerols.

   Phospholipid: a triglycerol in which one of the fatty acid groups is replaced by a phosphate group.

   Carbohydrate: made of the elements C, H and O, where H and O are present in the same proportions as in water. The basic unit here is a sugar (saccharide), which may be condensed with other sugar units.

5 e.g. Red cell: unable to grow, being reduced to a 'package' of haemoglobin and carbonic anhydrase enzyme.
   e.g. Relay neurones in brain: unable to store glycogen (as reserves for glucose).

3 Nutrition

1 (i) Saprotrophic nutrition, e.g. *Mucor* (pin mould);
   (ii) parasitic nutrition, e.g. *Taenia* (tapeworm).

2 Heterotrophs are dependent upon photosynthetic autotrophs for conversion of the Sun's energy into the chemical energy of sugars etc., and for supplying oxygen and removing carbon dioxide from the air.

3 The palate, partially separating the buccal cavity from the nasal cavity, permits breathing during mastication.

4 A vastly increased surface area of food permits more efficient enzyme action in the gut.

5 Starch, e.g. in bread and potatoes.

6 a) red kidney beans, peanuts, soya beans;
   b) cheese, meat, fish or eggs.

7 The source of the 20 or so amino acids required for protein synthesis. Proteins are needed in enzymes and cell membranes.

8 The villi of the small intestine vastly increase the surface area for the food to be exposed to digestive enzymes, and ultimately aid in absorption.

9 Lymph vessels from the small intestine drain into the blood circulation quite close to the heart, via the left subclavian vein.

10 See 'Organisation of the nervous system' (page 68).

11 The herbivorous mammal gains most of the products of cellulose digestion carried out by the bacteria, and the bacteria gain a habitat and food supply.

4 Exchange and transport

1 a) $C_6H_{12}O_6 + 6O_2 = 6CO_2 + 6H_2O +$ energy (much);
   b) $C_6H_{12}O_6 = 2C_2H_5COOH + 2CO_2 +$ energy (little).

2 Mucus (from goblet cells and mucous glands) traps dust from in-coming air. Cilia beat this stream of mucus away from lungs.

3 The concentration gradients of gases across the surface; the distance over which diffusion occurs; the area over which diffusion occurs; the chemical and physical structure of the diffusion surface.

4 See Table 5.2 (page 34).

5 Heart and circulation

1 An insect's blood is pumped at low pressure into spaces within the body, thereby bathing the organs.

2 A portal vein starts and ends in a capillary network, e.g. hepatic portal vein running from gut to liver.

3 Tissue fluid is an aqueous solution of the nutrients required by cells (oxygen, glucose, amino acids, fatty acids, inorganic acids), hormones, and waste products and excretory substances from cells (including carbon dioxide, urea); blood plasma also contains blood proteins and has blood cells suspended in it.

4 A 'coencytic' tissue lacks cell boundaries and appears as a multinucleate cytoplasm.

5 Brain cells have no store of glucose; they rely on a continuous supply of oxygen and glucose from oxygenated blood. Any brain cells destroyed in a stroke cannot be replaced.

6 Lysosomes are tiny organelles (page 7), formed from Golgi apparatus or from the endoplasmic reticulum. They contain hydrolytic enzymes that digest bacteria.

6 Excretion and osmoregulation

1 Animal cells do not have a cell wall to protect the plasma membrane. Instead, in mammals, the osmotic concentration of both cells and tissue fluids is osmoregulated (that is, regulation of the water potential of body fluids occurs by adjustment of water/salt concentrations). Excess salts and water may disrupt this regulation if they are not removed speedily.

2 Ammonia, a product of deamination of excess amino acids, is an extremely soluble and toxic substance. Urea is less toxic and less soluble, and therefore more safely disposed of than ammonia in relatively large terrestrial animals.

3 Na^+ and K^+ are involved in the action potential (pages 60–61), Ca^{2+} in muscle contraction (page 75) and bone formation (page 72), and Mg^{2+} is an enzyme co-factor for certain enzymes. NO_3^- is essential for green plants as it is used in amino acid synthesis; animals cannot manufacture their own amino acids.

7 Homeostasis

1 Liver cell organelles would include nucleus, mitochondria, Golgi apparatus, endoplasmic reticulum (rough and smooth), ribosomes and perhaps some lysosomes. These organelles are characteristic of cells with high metabolic activity.

2 With short-lived and frequently replaced enzymes, the cell, organ or organism can respond quickly to changing developmental instructions from the nucleus, or to changes in available substrates, or to changed physiological conditions.

3 In the complete respiration of glucose, glycolysis occurs in the cytoplasm, and the Krebs cycle and ATP synthesis occur in the mitochondria.

4 Endocrine glands secrete their products (hormones) directly into the bloodstream, whereas exocrine gland secretions pass out via a duct.

5 The essential components of a physiological feedback mechanism are: a detector device (some sense organ); a control centre (normally part of the brain) programmed with a set level of conditions, and effector systems (muscles or glands, perhaps), which can adjust the set condition back to the norm.

6 During sleep the body temperature normally falls below approximately 36.8 °C, and during physical activity or after consumption of a hot meal it tends to rise above this value.

7 The definition of a species implies 'organisms that are *not* capable of producing fertile offspring when mated with a different species'. We also use the term species exclusively for wild, not domesticated, organisms. Bactrian and dromedary camels are mainly domesticated and are known to interbreed to produce fertile offspring. Therefore, they should not strictly be classed as different species.

8 Sensitivity, coordination and control

1 Mitochondria are the site of ATP synthesis in aerobic respiration; the Golgi apparatus synthesises and packages vesicles of biochemicals.

2 Active transport requires energy from respiration and protein 'carrier' molecules; facilitated diffusion is diffusion aided by a molecule present in the membrane that is being crossed.

3 For a brief period following the passage of an action potential the axon is no longer excitable, whilst the resting potential is being re-established (called the refractory period).

4 Ethanol slows the metabolic activity of neurones, having a slightly anaesthetic effect. This is of special consequence to the fine control/judgement aspects of behaviour and body control.

5 The body requires constant feedback on its position and movements. Pressure receptors provide much of this.

6 Pupil size affects depth of focus; a reduction in the pupil aperture greatly increases depth of focus.

7 When the brain is also involved, the reflex arc involves relay neurone(s) in the grey matter in the spinal cord. When the brain is not involved, only sensor and motor neurones are essential components.

8 The 'flight or fight' response involves:
   (i) conversion of liver glycogen into glucose, increasing blood sugar levels;
   (ii) dilation of bronchioles and relaxation of smooth muscle of the gut, leading to a lowered diaphragm, more air inhalation and an enhanced supply of oxygen to respiring cells;
   (iii) increase in the delivery of blood to the tissues via increases in volume and rate of heartbeat and blood pressure;
   (iv) vasoconstriction of blood vessels to the gut and reproductive organs, and inhibition of peristalsis, thereby improving blood supply to the voluntary muscles;
   (v) decreased sensory threshold (heightened sensitivity) and increased mental awareness (more rapid responses);
   (vi) contraction of hair erector muscles, which gives an impression of large size in furry animals.

9 The blood capillaries supply oxygen, nutrients and hormones, and they remove waste products.

9 Support and locomotion

1 Heart and lungs (and liver, to some extent) are protected by the rib cage and sternum; brain, by the cranium; eyes, by the orbits.

2 Involuntary muscle occurs in sheets and bundles in the body wall and various internal organs; cardiac muscle occurs in the heart.

3 A multinucleate structure such as a voluntary muscle fibre is called a syncytium.

4 Heel raised/body swung forwards by contraction of (5); leg protracted (flexed) by contraction of (2), and flexed at knee by contraction of (3); leg now swings past other leg (being used as strut, i.e. taking weight of body), extended by relaxation of (3) and continuing contraction of (2), and with toes raised by contraction of (4); finally, mass of body swings over so that this leg starts to take the body's weight, after heel touches ground by contraction of (1).

10 Reproduction

1 Asexual reproduction involves splitting (by fission or fragmentation), spore formation or budding, to form new individuals. It does not involve the union of gametes (sex cells) to form a new individual. Asexual reproduction produces progeny that are genetically identical to the parent organism, whereas the progeny of sexual reproduction are similar but not identical to their parents.

2 A lower temperature than that of the body, normally about 2 °C lower, is required for sperm production.

3 Growth of the sex organs; growth of body hair (facial and pubic); enlargement of the larynx leading to a deeper voice; and general muscular development of the body.

4

Spermatogenesis	Oogenesis
Divisions of germinal epithelium commence at puberty and continue through adult life	Divisions of germinal epithelium occur in fetus (before birth) only
Spermatogonia themselves divide repeatedly, so that vast numbers of male gametes are formed. Both divisions of meiosis occur in the testes. Has occurred before fertilisation	Oogonia surrounded by follicle cells form primary follicles that remain dormant until puberty i.e. do not subsequently divide mitotically. Meiosis I occurs in the ovaries, but meiosis II is not completed (in oviduct) until fertilisation occurs
Meiosis produces equal-sized cells	Meiotic divisions produce secondary oocytes and polar bodies (i.e. unequal size)

5

Structure	Endocrine output	Main role in regulating menstrual cycle
Anterior lobe of pituitary	Secretes FSH and maintains LH output	Activates a few follicles to start to develop
Wall of dominant follicle	Secretes oestrogen; then, when maximum oestrogen concentration is reached, secretes oestrogen + FSH + LH	Starts by inhibiting FSH production; stimulates surge in FSH + LH production; triggers ovulation
Corpus luteum (vacant follicle)	Secretes progesterone	Oestrogen and progesterone (i) inhibit FSH + LH production, and (ii) maintain endometrium
	Falling levels of FSH + LH	Allow corpus luteum to degenerate; endometrium breaks down

6

Roles of the placenta	How these occur/are carried out
Supply of O_2	Diffusion between maternal and fetal blood
Removal of waste products of metabolism	Active transport and diffusion between maternal and fetal blood
Nutrients (sugars, amino acids vitamins, etc.)	Active transport and diffusion between maternal and fetal blood
Secretes HCG hormone (takes over this role from blastomeres)	Maintains corpus luteum, which maintains the endometrium for first 16 weeks

Glossary

Entries are *aides-mémoire*, rather than formal definitions.

action potential   An impulse; a rapid change in membrane potential of an excitable cell, e.g. a neurone.

active transport   Movement of substances across a membrane involving a carrier protein and energy from respiration.

adenosine triphosphate (ATP)   A nucleotide present in every living cell, formed in respiration from ADP and Pi; functions in metabolism as a common intermediate between energy-requiring and energy-yielding reactions.

adrenaline   A hormone secreted by the adrenal medulla (and a neurotransmitter secreted by nerve endings of the sympathetic nervous system). Has many effects, including speeding of heartbeat, and the breakdown of glycogen to glucose in muscle and liver.

aerobic respiration   Respiration requiring oxygen, involving oxidation of glucose to CO_2 and H_2O.

antibody   A protein produced by blood plasma cells. Derived from B-lymphocytes when in the presence of a specific antigen; binds with the antigen and aids its destruction.

antidiuretic hormone (ADH)   A hormone secreted by the pituitary gland; controls the permeability of the walls of the collecting ducts in the kidney.

atherosclerosis   Deposition of plaque (cholesterol derivative) in the inner wall of blood vessels.

body mass index (BMI)   (Body mass (kg)/Height (m))^2.

brush border   Tiny, finger-like projections (microvilli) on the surface of epithelial cells of the ileum; facilitate absorption of digested food.

cardiac cycle   The stages of the heartbeat. The atrium walls and then the ventricle walls alternately contract (systole), followed by relaxation (diastole).

carrier protein   One of the types of protein in plasma membranes; responsible for active transport across the membranes.

cholesterol   A lipid found in animal plasma membranes; a precursor of the steroid hormones. In humans it is formed in the liver and transported in the blood as lipoprotein.

chromosome   Visible (in appropriately stained cells) at nuclear division, appearing already replicated as chromatids. Each chromosome/chromatid consists of a long thread of DNA packaged with a protein (histone). At other times, it may appear as chromatin.

cleavage   A series of divisions in which the zygote is transformed into a ball of cells.

corpus luteum   A glandular mass that develops in mammals after the ovum is discharged from an ovarian follicle.

cytoplasm   The living part of the cell bound by the plasma membrane; excluding the nucleus, it consists of organelles and cytosol.

deamination   The removal of NH_2 from an amino acid, e.g. as a prelude to urea formation.

dendrite   A fine fibrous process on a neurone that receives impulses from other neurones.

deoxyribonucleic acid (DNA)   A form of nucleic acid found in chromosomes in the nucleus; consists of two complementary chains of deoxyribonucleotide sub-units, and contains the bases adenine, thymine, guanine and cytosine.

depolarisation (of axon)   A temporary and local reversal of the resting potential difference of the membrane that occurs when an impulse is transmitted along the axon.

diffusion   The net movement of atoms, ions or molecules from a region of their high concentration to a region of their low concentration due to kinetic energy.

double circulation   A circulation in which the blood passes through the heart twice (pulmonary circulation then systemic circulation) in any one complete circuit of the body.

effector   A cell or organ that responds to an impulse by doing something, e.g. a muscle contracts, a gland secretes.

endocrine glands   The hormone-producing glands (ductless) that release their secretion directly into the body fluids.

endothermic   Generation of body heat metabolically.

facilitated diffusion   Diffusion across a membrane, facilitated by molecules in the membrane (but without the expenditure of metabolic energy).

fluid mosaic model   The accepted view of the structure of the plasma membrane: a phospholipid bilayer with proteins embedded but free to move about.

gaseous exchange   Exchange of respiratory gases (O_2, CO_2) between cells/organism and the environment.

gastrulation   The early stage in embryology involving cell movements, formation of the gut and laying down the primary layers (ectoderm, mesoderm, endoderm).

generator potential   The localised depolarisation of the membrane of a sensory cell.

gestation   Time between fertilisation and birth in a viviparous animal.

glycolysis   The first stage of tissue respiration in which glucose is broken down to pyruvic acid, without the use of oxygen; occurs in the cytosol.

gonadotrophic hormone   follicle-stimulating hormone (FSH) and luteinising hormone (LH); secreted by the anterior pituitary; stimulates gonad function.

habituation   Adjustments by which contact with the same stimulus produces a diminished effect.

haemoglobin   A conjugated protein found in red cells; effective at carrying oxygen from regions of high partial pressure (e.g. lungs) to regions of low partial pressure (e.g. respiring tissues).

homeostasis   The maintenance of a constant internal environment.

hydrolysis   A reaction in which hydrogen and hydroxyl ions from water are added to a large molecule causing it to split into smaller molecules.

hyperglycemia   Excess glucose in the blood.

hypoglycemia   Very low levels of blood glucose.

hypothalamus   A part of the floor at the rear of the forebrain; a control centre for the autonomic nervous system, and a source of 'releasing factors' for pituitary hormones.

implantation   Embedding of the blastocyst (developed from the fertilised ovum) into the uterus wall.

inhibitory synapse   A synapse at which arrival of an impulse blocks the forward transmission of impulses in the post-synaptic membrane.

inspiratory capacity   The amount of air that can be drawn into the lungs.

isotonic   Being of the same osmotic concentration, and therefore, of the same water potential.

lactation   Secretion of milk from mammary glands.

loop of Henle   The loop of the mammalian kidney tubule, passing from cortex to medulla and back; important in the process of urine concentration.

lysis   Breakdown, typically of cells.

meiosis   Nuclear division in which the daughter cells contain half the number of chromosomes of the parent cell.

menopause   Cessation of ovulation and menstruation in women.

menstrual cycle   Monthly cycle of ovulation and menstruation in human females.

mRNA   Single-stranded ribonucleic acid formed by the process of transcription of the genetic code in the nucleus; moves to the ribosomes in the cytoplasm.

mitosis   Nuclear division in which daughter nuclei have the same number of chromosomes as the parent cell.

motor neurone   Nerve cells that carry impulses away from the central nervous system to an effector (e.g. muscle, gland).

myelin sheath   An insulating sheath of axons formed by Schwann cells, which wrap their cell bodies around nerve fibres.

myogenic   Originating in heart muscle cells themselves, generating the basic heartbeat.

neurotransmitter substance   A chemical released at the pre-synaptic membrane of an axon upon arrival of an impulse; it transmits the 'impulse' across the synapse.

nuclear division   The first step in the division of a cell, when the contents of the nucleus are subdivided by mitosis or meiosis.

oestrous   The period of fertility (immediately after ovulation) during the oestrous cycle.

oogenesis   The sequence of formation of the ovum from the germinal epithelium, which occurs in ovary and oviduct.

organogeny   The formation of organs during embryology.

osmoregulation   Regulation of the water potential of body fluids by the regulation of water and/or salt content.

ovarian cycle   The monthly changes that occur to ovarian follicles leading to ovulation and the formation of a corpus luteum.

ovulation   Shedding of ova from the ovary.

pacemaker   The area of origin of the myogenic heartbeat, known as the sino-atrial node.

perception   The mental interpretation of sense data (i.e. occurring in the brain).

peristalsis   The wave of muscular contractions passing down the gut wall.

phagocytosis   Ingestion of solid matter by cells, by invagination of the plasma membrane to form a food vacuole.

pinocytosis   Uptake of a droplet of liquid into a cell involving invagination of the plasma membrane.

potential difference   The separation of electrical charge within or across a structure, e.g. a membrane.

reductive division   Meiosis in which the chromosome number of a diploid cell is exactly halved.

reflex action   A response automatically elicited by a stimulus.

reflex arc   A functional unit in the nervous system; consists of sensory receptor, sensory neurone, (possibly relay neurones), motor neurone and effector (e.g. muscle or gland).

refractory period   The period after excitation of a neurone, when a repetition of the stimulus fails to induce the same response; divided into periods known as absolute (initially), and relative (later).

relay neurone   Nerve cell of the grey matter of the CNS, functioning very much as the name suggests.

resting potential   The potential difference across the membrane of a neurone when it is not being stimulated.

ribonucleic acid (RNA)   A form of nucleic acid containing the pentose sugar ribose, found in the nucleus and cytoplasm of eukaryotic cells (and sometimes the only nucleic acid of prokaryotes). It contains the organic bases adenine, guanine, uracil and cytosine.

saltatory conduction   When an impulse conducts 'in jumps', between nodes of Ranvier.

sensory neurone   Nerve cell that carries impulses from a sense organ towards the CNS.

sensory receptor   A cell specialised to respond to stimulation, by the production of an action potential.

sexual reproduction   Involves the production and fusion of gametes.

spermatogenesis   The sequence of formation of sperm from germinal epithelium, which occurs in the testes.

stimulus (plural stimuli)   A change detected by the body that leads to a response.

summation   The combined effect of many nerve impulses; can either be spatial (from different axons) or temporal (via a single axon).

synapse   The connection between two nerve cells; functionally, a tiny gap called the synaptic cleft is traversed by transmitter substances.

target organ   The organ on which a hormone acts (although it broadcasts to all organs via the blood).

threshold of stimulation   The level of stimulation required to trigger an action potential (impulse).

thrombosis   Blood clot formation, leading to the blockage of a blood vessel.

tidal volume   Volume of air normally exchanged in breathing.

tissue respiration   The biochemical steps by which energy is released from sugars.

turgid   Having high internal pressure.

Index